DISCARD

More Praise for *The Six*

"Propulsive, startling, and deeply moving, *The Six* captures the odyssey of America's first women astronauts, tracking them in delightful detail from the ground to the sky, from earthbound battles against sexism to stunning feats performed in literal orbit. Loren Grush is masterful."

—Jason Fagone, bestselling author of *The Woman Who Smashed Codes: A True Story of Love, Spies, and the Unlikely Heroine Who Outwitted America's Enemies*

"Compellingly intimate portraits of a group of brave astronauts who changed the face of NASA—and exploration . . . Well-researched and gripping, *The Six* is an inspiring and, at times, maddening tale that reminds us what the definition of hero is."

—Christian Davenport, Peabody Award–winning staff writer at the *Washington Post* and author of *The Space Barons*

"Today there is nothing unusual about a woman flying in space, walking in space, or living in space, which makes it difficult to imagine what it was like forty-five years ago for the six women who broke the highest of all glass ceilings to become astronauts. As Loren Grush shows in this illuminating book, they overcame daunting obstacles to make their indelible marks on Earth and in space."

—Andrew Chaikin, author of *A Man on the Moon: The Voyages of the Apollo Astronauts*

"Loren Grush's *The Six* tells this triumphant and sometimes tragic story as skillfully and completely as it ever will be told, dramatically showing events through the eyes and ears of the women themselves."

—Michael Cassutt, author of *The Astronaut Maker: How One Mysterious Engineer Ran Human Spaceflight for a Generation*

"Recreates the crucial turning point when America turned toward spaceflight equality, inviting six brave and brilliant women to become NASA's first female astronauts . . . Compelling."

—Eric Berger, senior space editor, *Ars Technica*, and author of *Liftoff*

THE
SIX

The Untold Story of America's First Women Astronauts

Loren Grush

SCRIBNER

New Delhi

Scribner
An Imprint of Simon & Schuster, Inc.
1230 Avenue of the Americas
New York, NY 10020

First Scribner hardcover edition September 2023

SCRIBNER and design are registered trademarks of The Gale Group, Inc.,
used under license by Simon & Schuster, Inc., the publisher of this work.

For information about special discounts for bulk purchases,
please contact Simon & Schuster Special Sales at 1-866-506-1949 or
business@simonandschuster.com.

The Simon & Schuster Speakers Bureau can bring authors to your live event. For
more information or to book an event, contact the Simon & Schuster Speakers
Bureau at 1-866-248-3049 or visit our website at www.simonspeakers.com.

Interior design by Davina Mock-Maniscalco

Manufactured in the United States of America

10 9 8 7 6 5 4 3 2 1

Library of Congress Cataloging-in-Publication Data has been applied for.

ISBN 978-1-9821-7280-0
ISBN 978-1-9821-7282-4 (ebook)

To the women who look up and dream of more.

CONTENTS

CONTENTS

Sally Ride

Born May 26, 1951. Hometown: Encino and Van Nuys, California. PhD, master's, and bachelor's in physics, bachelor's in English, Stanford; junior tennis champion. Accepted as NASA astronaut 1978. First American woman to fly in space. Veteran of two Shuttle missions. Member of the Rogers Commission investigating the *Challenger* explosion, as well as the *Columbia* Accident Investigation Board that investigated the *Columbia* disaster.

Judy Resnik

Born April 5, 1949. Hometown: Akron, Ohio. PhD in electrical engineering, University of Maryland; bachelor's in electrical engineering, Carnegie Mellon. Accepted as NASA astronaut 1978. Second American woman to fly in space, and first Jewish American to fly in space. Died while flying aboard *Challenger*.

Kathy Sullivan

Born October 3, 1951. Hometown: born in Paterson, New Jersey, grew up in Woodland Hills, California. PhD in geology, Dalhousie University; bachelor's in Earth sciences, University of California,

Santa Cruz. Accepted as NASA astronaut 1978. Third American woman to fly in space. First American woman to walk in space. Veteran of four Shuttle missions, including the launch of the Hubble Space Telescope. The only person to have walked in space and traveled to the deepest part of the ocean.

Anna Fisher

Born August 24, 1949. Hometown: born in St. Albans, New York, grew up in San Pedro, California. MD, UCLA; master's and bachelor's in chemistry, UCLA. Accepted as NASA astronaut 1978. Fourth American woman to fly in space, and first mother to fly in space. Flew one Shuttle mission.

Margaret "Rhea" Seddon

Born November 8, 1947. Hometown: Murfreesboro, Tennessee. MD, University of Tennessee, bachelor's in physiology, UC Berkeley. Accepted as NASA astronaut in 1978. Fifth American woman to fly in space. Veteran of three Shuttle flights.

Shannon Lucid

Born January 14, 1943. Hometown: born in Shanghai, China, grew up in Bethany, Oklahoma. PhD and master's in biochemistry, University of Oklahoma, bachelor's in chemistry, University of Oklahoma. Accepted as NASA astronaut in 1978. Sixth American woman to fly in space. Veteran of five spaceflights. At one time held the record for longest continuous stay in space by an American and by a woman.

nna Fisher sat alone inside the dim cockpit of the space shuttle *Challenger*, the dark midnight sky painted across the cabin's glass windows above her. The cramped room, often the setting of vibrant chatter and activity, stood still and quiet that night. The analog screens on the main control panel displayed only gray. Thousands of metallic switches, hand levers, and multicolored toggles blanketed the walls and ceiling around her—all motionless.

In the dark she gazed at the little devices intently, making sure they stayed right where they were. All the switches had been configured in just the right positions for when morning came: when the Space Shuttle would ignite its three main engines, generating more than a million pounds of thrust and rocketing a crew of five into the heavens.

Clad in a flight suit—standard issue for a NASA astronaut— Anna, a medical doctor by training, reclined on her back, stretched out in one of the cockpit's chairs. With the Shuttle fully upright on its launchpad, she loomed high above the Central Florida coastline. The cockpit put her roughly eighteen stories up, though it was hard to tell in the dark. The only indication of the height came from the eerie light filtering in from outside. Bright xenon floodlights surrounding the launchpad bathed the spacecraft in a ghostly glow that seeped into the cockpit's interior.

Anna may have appeared to be the cabin's lone occupant that night, but she actually had some company. A large belly bump pushed up against her suit as she lay in the darkness. Just one month away from giving birth, Anna's soon-to-be daughter, Kristin, was also present. The sight of a pregnant astronaut was an unusual one for the Shuttle cockpit. Anna was just the second in the corps to become pregnant in the space program's history. For decades prior, such a thing hadn't been possible. The astronauts had all been men.

Anna's job was simple that evening: maintain the status quo. Earlier that day, other astronauts and flight personnel had been inside the cockpit, flipping the more than two thousand switches and buttons to their current settings. After the teams had finished their task, Anna took over to keep a watchful eye on the cabin. She was there to ensure no switch got accidentally flipped to the wrong position, a tiny error that could jeopardize the launch. Anna wouldn't be flying in the morning, but she'd still be key to avoiding a last-minute delay or abort. It wasn't a glamorous task, but it was crucial.

Plus, Anna loved it. After five years of sometimes grueling training, this night's assignment was a fun one—a tiny dress rehearsal for when she'd fly on the Space Shuttle herself. She was also demonstrating her full commitment to the job, proving to any lingering doubters that her pregnancy wouldn't get in the way of her astronaut duties. Most of all, she was doing her part to guarantee the success of the mission on deck, a flight that would become one of the most consequential missions in NASA history.

The night was June 17, 1983. The next day, fellow astronaut Sally Ride would become the first American woman to fly to space.

The late-night hours transitioned to early-morning ones, and Anna eventually left the cockpit to make way for the crew. Once she departed *Challenger*, flight engineers began pumping more than 500,000 gallons of freezing liquid propellants into the spacecraft's massive external tank—a bulbous burnt-orange structure strapped to the Shuttle's underbelly. Picked for their propensity to combust when combined, the super-chilled liquids would sit inside the tank, waiting to come together and ignite, at which point the vehicle would be lofted into the sky. NASA's giant countdown clock, located on a grassy field a few miles from the launchpad, ticked away to that exact moment.

A few more miles away, a small, curly-haired brunette stepped through the open gray metal doors of the astronaut quarters and out into the pre-dawn nighttime. Surrounded by her four male crewmates, astrophysicist Sally Ride, wearing a light blue onesie flight suit, descended the ramp out of the building, smiling and lifting her right hand to give a slight wave to a nearby crowd of photographers. Her face didn't betray any hint of nervousness. Resolute, she walked forward with her crewmates, before turning a corner. Then, one by one, they climbed into an off-white Itasca Suncruiser RV with a horizontal brown stripe cutting across its sides. Among those accompanying the crew was George Abbey, a well-built middle-aged man with a crew cut and wearing a dark suit. George had selected Sally and her crew to fly on this mission, and he was there to see the astronauts one last time before they broke free of Earth's gravity.

With the stars still twinkling above, the van sped down a lone concrete road to its destination: the launchpad at LC-39A. The site had been home to previous trailblazing launches. It was the same place where astronauts Neil Armstrong, Buzz Aldrin, and Michael Collins had taken off in a monster Saturn V rocket more than a decade earlier, bound for the moon. Now, the launchpad was ready to host another pivotal mission—one that many felt was coming much too late.

The RV carrying Sally's crew slowed to a halt. They idled just three miles from the pad, stopping right outside the famous Launch Control Center, a squat gray and white building with thick concrete walls and wide sweeping windows angled toward the sky. Since the Apollo program, the Launch Control Center had served as the central hub for monitoring all human spaceflight missions out of the Cape. It was no different today. Once more, in those early-morning hours, the facility was buzzing with energy and nervous excitement as everyone inside prepared for the mission at hand. This was where George got out of the vehicle. He wished the crew a good flight and headed toward the center. The van drove onward, LC-39A in its sights.

George took his place inside the LCC's firing room, where hundreds of workers sat in front of gray consoles, intently watching their screens. Overhead, wide slanted windows loomed, providing a tantalizing reason to intermittently gaze outside. The glass provided a panoramic view of the launchpad in the distance—perhaps the best view one could get of a spacecraft taking off from LC-39A. The only location that might rival it was the roof of the control center. That's where Anna would be heading shortly—to watch the Shuttle take off.

The Suncruiser finally pulled up to the pad, and Sally stepped out. She and her crewmates walked toward the giant black-and-white Space Shuttle standing triumphantly in front of them. Two towering white rocket boosters the size of skyscrapers stood attached to the massive orange tank on the Shuttle's underside, poised to ignite at liftoff and provide the millions of pounds of thrust needed to propel the crew off the ground. With the Shuttle system fully assembled in front of her, Sally didn't feel like she was standing before an inanimate spacecraft, but rather a monstrous breathing animal. The Shuttle's liquid propellants periodically vented like steam from a kettle, causing the vehicle to hiss and moan as if it were alive. It took all of Sally's will to focus on just moving forward.

The five crewmates loaded into an elevator near the Shuttle's base and found themselves ensconced in a large metallic service structure. One of them pushed a button, starting their climb up to the 195-foot level. Reaching the top, they stepped from behind the elevator doors and trekked across a long suspended walkway jutting out into the Florida air. At the end of the corridor stood a small white room, which displayed the open hatch of the Space Shuttle cockpit. Sally and the crew gathered in the room, where a team of flight technicians were waiting for them. Sally donned a cap, her radio headset, and helmet. Then, individually, she and her fellow astronauts entered the Shuttle to be strapped into their seats for the ride ahead. Sally took her place in the seat directly behind the pilot and the commander, a spot that would give her a direct view of the control panels and the Space Shuttle's windows during the ascent. She sat still as the closeout crew strapped her into the hard metal seat.

And then she waited, lying on her back, just as Anna had done earlier that night. With the crewmates strapped in, the closeout team left the cockpit and shut the hatch. *Oh my gosh, this is really going to happen*, Sally thought.

At around 6:30 a.m., the hot Florida sun peaked out from beneath the horizon, pulling high above the nearby Atlantic Ocean. The sun's rays sizzled on the sandy beaches and swampy, gator-filled marshlands surrounding Cape Canaveral, Florida—home of NASA's Kennedy Space Center and the Space Shuttle's primary launch site. In the dawn light, giant throngs of people could be seen camped out with tents and folding chairs along the various shorelines of the Central Florida coast. Thousands more stood and waited along the roadways of the nearby sleepy beach town of Cocoa Beach, wearing shorts, tank tops, and visors to combat the crushing heat. A few vendors sold T-shirts while music blared from handheld radios. The song "Mustang Sally" could be heard blasting across the coast. One sign at a local bank branch read: RIDE, SALLY RIDE! AND YOU GUYS CAN TAG ALONG, TOO!

Potentially, a half-million people would be putting eyes on the Shuttle that day. And that didn't count the possibly millions more watching on their TV sets. Many of the in-person gawkers had flown or driven hundreds of miles to make it to the edge of the United States for this launch, just to catch a tiny glimpse of history being made.

Later that morning, another astronaut stepped out into the Florida sun just a few miles from the launchpad. Shannon Lucid made her way to a quiet spot on a beach near the space center, far from the bustling crowds that didn't have the privilege of coming onto NASA grounds. She'd been up late the night before, also working inside *Challenger's* cockpit, checking all the switches that Anna had then been tasked with monitoring overnight. For this mission both Shannon and Anna were "Cape Crusaders," the informal job title given to the astronauts who helped support the flight. Anna had been the *lead* Cape Crusader this round. And one of the perks of the job was getting to watch a launch alone on a relatively private beach in peaceful silence. All Shannon could hear that morning were the waves lapping the sandy shore.

She didn't much care that she wasn't the one sitting in Sally's seat that morning. Just having a job—let alone one as unique and daring as an astronaut—meant the world to her. Before she'd come to NASA, she'd struggled to find employers who wouldn't judge her for her status as both a woman and a mother. But with the space program, she'd finally found a place that would allow her to work as the individual she'd become, while putting her chemistry and science expertise to use. And as long as she got to fly, it didn't matter who went first. Shannon waited as the countdown clock ticked away.

Standing beneath the open Florida sky, Anna looked out at the pad from the top of the Launch Control Center. The Shuttle that had been towering over her hours earlier now looked like a mere bump just along the horizon, nestled among the green treetops that spread throughout the Kennedy Space Center. A large crowd had started to form on the roof with her. A tall, lanky astronaut with bright red hair stood nearby: Sally's husband, Steve Hawley. Next to him stood Carolyn Huntoon, a friend and mentor of Sally's at NASA who'd taken the fledgling astronaut under her wing when she'd first been chosen to join the space program. They all chattered among themselves, swapping anxious and excited energy as they waited for the countdown to move into the final sequence.

As the seconds ticked by, more people inserted themselves into the crowd on the roof that morning. Rhea Seddon, a petite Tennessee native with light blond hair that barely brushed her shoulders, made her way to the roof's edge. It was also *her* first time watching a launch on the roof of the control center. Rhea had come to see her colleague take flight, eager to be sitting where Sally was now. She didn't know when her time would come but she was ready. A trained surgeon, Rhea had been splitting her time between emergency room work in Houston and her technical assignments at the space agency. She was also a new mom, having given birth to her son Paul less than a year earlier. She'd been the first astronaut to ever give birth, and she felt deep inside that she was ready for motherhood, her medical work, and a mission. She just needed the assignment.

It's possible that not far from Rhea stood Judy Resnik, tan with raven hair that fell in large curls around her face. She might have looked out on the horizon at the tiny Space Shuttle, dreaming about the moment when she'd be in the same place as Sally. Or she might have been watching the launch from a television screen. Her location that day isn't certain. But for Judy, an electrical engineer, there was no uncertainty about when *her* time would come. In George's office in February she'd been assigned to her own shuttle flight, an appointment that would make her the second American woman to fly after Sally. As the seconds passed by, she knew that her own moment was on its way, even if it meant she wouldn't be the first. To her it didn't matter. She just wanted to fly as quickly as possible. But when she *was* hurtling through the sky it would mean freedom, independence, and purpose—everything she'd spent her life searching for.

While it seemed that almost everyone was at the Cape that day, one key astronaut was missing. More than two thousand miles away,

Kathy Sullivan wasn't following the launch countdown but was instead running through a checkout list for a scuba dive at the Scripps Institution of Oceanography in San Diego, California. An avid explorer and seafarer, Kathy had always yearned to explore the world, whether it was from the vantage of the ocean's depths or from the towering height of the Shuttle's position above Earth. She planned this specific dive to complete her certification for open-water scuba diving before giving a commencement address at the University of California at San Diego the next day.

For her, it was a convenient way to escape the launch. She'd hoped *she* might be the one chosen to be the first American woman in space. Denied that honor, she decided to accept the invitation from UC San Diego. That way she didn't have to put on a happy face while watching the flight from Florida or Houston. It was the first Shuttle launch Kathy would miss seeing in person. She couldn't have known at the time that her own mission was not far off—that she'd be making history in another momentous way in just over a year.

Though they weren't all *physically* in the same place that day, Sally, Judy, Kathy, Anna, Rhea, and Shannon were connected in a way that would span distances and decades. They were the Six: the first six women astronauts NASA had ever chosen, after years in which the space program had only picked men to inhabit the spaceships that rocketed off Earth. Selected as mission specialists to fly on the Space Shuttle in 1978, they'd go on to become the first six American women to fly into the dark void of space, earning the unique privilege of seeing the curvature of the Earth reflected in their eyes. They'd deploy satellites and telescopes, cope with microgravity in spacesuits, maneuver robotic arms, meet presidents and royal dignitaries, speak in front of thousands of wannabe space explorers, and pave the way for dozens of women to come after.

While all earned the status of pioneers simply by being selected, one had to go first and break the country's highest glass ceiling before the rest. Sally had been that one, achieving a feat that would make her a household name alongside the likes of Alan Shepard, John Glenn, and Neil Armstrong. Being tapped to blaze the trail before anyone else would turn her into a feminist icon, a role model, and part of a Billy Joel lyric. It would also place a burden on her shoulders that she'd carry the rest of her life.

But her assignment hadn't been guaranteed from the start. The designation could have easily gone to Judy or Anna or any of the other women waiting in line. All were thoroughly qualified. It took a final judgment call from George Abbey and the NASA management team to seal the fates of the Six, putting Sally at the head of the pack and the others right behind her. The subsequent flight assignments would irrevocably alter the rest of their lives, leading to dreams fulfilled, historic firsts, and unimaginable tragedy.

George's fateful nod led to Sally's lying on her back inside the space shuttle *Challenger* just after 7:30 a.m. on the morning of June 18, 1983. She stared ahead, thinking about the job in front of her and how she just wanted to do it right. After three hours of reclining horizontally on the hard gray metal seat, she heard the flight controller in her ear count down to the moment that would change everything.

"T-minus ten. Nine. Eight. Seven. Six. We are 'go' for main engine start. We have main engine start. And ignition . . ."

CHAPTER 1

But Only Men Can Be Astronauts

The sun was still hours away from coming up that morning, and Margaret "Rhea" Seddon was already staring into the open abdomen of a patient on her operating table. As per usual, she was trying to control the patient's blood flow and repair the damage to his organs, caused by a bullet that had violently ripped through his gut. By now, she'd become used to such grisly sights. As a surgical resident in John Gaston Hospital's emergency room, she saw all manner of gruesome gunshot and knife injuries, often the result of two angry men and too much beer. EMS crews would wheel into the John—as the doctors called their hospital—the victims of these bar brawls, typically in the middle of the night, and they'd become Rhea or some other doctor's priority for the rest of the evening. Each night, the John's emergency room would see so many trauma patients that it would earn an even more menacing nickname: the pit.

Sometimes Rhea stanched the blood and sewed people up just fine; other times she just couldn't repair the sheer amount of damage. Those moments were the most devastating. This early morning, however, things seemed to be progressing well, and she eventually stitched up her patient, sending him off to the ICU. Her work wasn't done, though. As the patient's doctor, she still had to keep her eye on him in case some unforeseen complication popped up. So she headed to the doctors' lounge, which sat adjacent to the ICU.

It was hard for her to believe, but there'd been a time when her presence in the doctors' lounge would have been a serious transgression. When she'd been a surgical intern at Baptist Memorial Hospital, across the street from the John in Memphis, Tennessee, she'd been barred from entering the doctors' lounge. It was for "men only," and she was the only woman surgical intern at the time. The head doctor told her the reason was that sometimes men walked around in their underwear in the lounge. She told him it didn't bother her, but he said the men would be embarrassed. Her superiors told her she could wait between surgeries in the nurses' bathroom. Rhea tried to change the policy, but she lost out and found herself taking naps on a foldout chair in the bathroom, with her head resting against the wall. The rule had prompted her to switch to the John for her residency—a place that didn't cling to such sexist policies.

Ever since she'd decided to go into medicine, Rhea always seemed to be out of her comfort zone in some way. She'd grown up in a completely different world: a small girl with straight blond hair in the upper-middle-class suburban town of Murfreesboro, Tennessee. There, she followed the standard recipe for How to Make a Proper Southern Lady. She took the requisite ballet lessons from Miss Mitwidie's Dance Studio. She learned formal dining etiquette, played the piano, sewed buttons on dresses, and planted herbs. Those skills were the ones her mother, Clayton, had learned in *her* youth, and she was simply passing the torch on to her daughter, molding Rhea into the only type of girl she knew how to make.

"People always followed in their parents' footsteps, and I always thought I would be like my mother—be a southern belle and stay home and cook and raise babies," Rhea said. But her father, Edward, had other ideas. An attorney, Edward wanted Rhea to have more than what her mother and grandmother had growing up. That meant exposing Rhea to a more diverse range of experiences. One night in October 1957, Edward pulled Rhea outside and pointed her gaze toward the darkened sky. There she watched a tiny blip of light zoom through the darkness. The tiny dot was Sputnik, an aluminum-based satellite the size of a beach ball that was beeping as it circled the Earth. The Soviet Union had just launched the space-craft a day or two prior on October 4, putting the first human-made object into orbit. Fear had coursed through the American public over the Soviet Union's newfound space dominance.

But others also realized that it was a watershed moment for *everyone*, not just the Soviets. "You are watching the beginning of a new era," Edward said. "It's called the Space Age." Although she was a month shy of turning ten years old at the time, Rhea's still-forming mind could grasp that a new world was on its way. However, she didn't quite realize at the time just how big a role space would play in her life.

The launch of Sputnik would ultimately put Rhea on a different path than the prim and proper one her mother had envisioned for her. One of America's knee-jerk responses to Sputnik was to increase the level of science education in grade schools, an attempt to train the next generation of youngsters to become a new breed of brainiac who could keep the US competitive in the unfolding Space Race. Soon, Rhea fell in love with her science courses—particularly life sciences. But she still stuck to the recipe. She would eagerly dig through the innards of a dissected rat during the day while performing stunts with the cheerleading squad after school.

For undergrad, she journeyed to the University of California at Berkeley, a place she'd chosen for its life sciences program. But the school felt as if it existed on another planet, let alone another state. She started as a freshman in 1965, and arrived at a campus bursting with political activism. The year prior to her arrival the Free Speech Movement had erupted when students flooded the university, protesting the arrest of a fellow student for handing out leaflets on campus. An era of political protests followed—with some students decrying the atrocities of the Vietnam War and others championing the civil rights movement and the Black Panthers. The heated rhetoric and liberal tilt were a shock for a seventeen-year-old from a conservative Tennessee town that had a third of UC Berkeley's population.

Rhea's GPA struggled that first year. Then the summer after her freshman year, she got a taste of a surgeon's life. Her father, Edward, had been on the board of directors of a small hospital in her hometown, and he arranged for Rhea to get a summer job there. Originally, she'd planned to work in the ICU of the hospital's new Coronary Care Unit, but its opening had been delayed. Instead, the doctors sent her to work in surgery. Rhea was hooked from the start. She'd leveled up from peering into the open cavities of dissected frogs and rodents; now, she was staring inside the stomachs of actual patients.

That job guided her when she entered medical school at the University of Tennessee in 1970, since, by then, she knew she wanted to follow the path of a surgeon. It was a path that almost didn't happen. In college she'd almost gotten married, and even had a wedding date set. "Came close to the time of the wedding, and I said, 'Not going to work,'" Rhea would later say. "He wants me to iron his shirts and stay at home and not go to work. So I backed out of the wedding." In hindsight, it turned out to be the right decision.

In her first year, she'd been one of just six women in a class of more than one hundred. As she worked her way through school, her internship, and her residency, she got used to being surrounded by men. And while the idea of becoming a doctor was top of mind, she also had a hidden motive for going to medical school. Secretly, she wondered if it might lead to a future in space. In watching Sputnik career across the sky, a seed had been planted. She figured that one day there'd be a future with space stations orbiting Earth, staffed with doctors. Perhaps *she* could be one of those doctors to live in space.

It had been the most minuscule thought, but it had stayed with her for years. And it was on her mind that early morning in the doctors' lounge after the gunshot wound surgery. Smelling of body odor and other human fluids, Rhea contemplated if this was the life she really wanted. At that moment a neurosurgery resident named Russ waltzed into the lounge and sat next to her. He seemed to be having the same existential crisis as Rhea, and the two friends commiserated over their exhaustion. He then posed a question that many ask when their current situation seems dim.

"What would you do if you weren't doing this?"

Rhea paused for a moment and then answered honestly. "I'd be an astronaut." It was perhaps the first time she'd ever said anything like that.

To anyone else it may have been a strange response, but Russ had a surprising reply. "I used to work for NASA," he said. He explained his old job, though Rhea didn't quite fully understand what he did. But Russ noted that he still kept in touch with his former coworkers in the space program. The conversation petered out from there, and the two residents eventually got up from their seats to finish their rounds.

A couple weeks later, Rhea was back at the John, going through her typical day of opening human bodies and stitching them shut. At one point she saw Russ, who was rushing to his next surgery. As the two passed each other, he suddenly remembered what Rhea had told him in the doctors' lounge and stopped her. "Hey, some friends of mine say they're taking applicants for the Space Shuttle program," he said. "I hear they have an affirmative action program!" He then hurried off without giving any more information.

Rhea stood there stunned, a million questions running through her head.

IT WAS AUGUST 1963 and Shannon Wells could almost taste freedom. In just two weeks, she was about to graduate from the University of Oklahoma with a bachelor's degree in chemistry. No more tests, no more papers. It was time to start thinking about actual employment. But as graduation day quickly approached, Shannon realized she didn't know anyone in her life who'd successfully snagged a job related to her field, and she had no idea how to get started. She figured one of her teachers might help her.

"I'm graduating in two weeks," Shannon recalled saying to her inorganic chemistry professor during one of her last classes. "How do I go about getting a job in chemistry?"

The professor seemed stunned. "What?" he asked, incredulous. "You're going to get a *job*?"

"Yes!" Shannon replied. "That's why I majored in chemistry."

Shaking his head, her teacher made his opinion clear: "There's absolutely *no one* who will hire you." He didn't say it outright, but Shannon understood what he meant. It was because she was a woman.

Shannon left the meeting shaken. But the hard truth was that her teacher had been right.

Sadly, dealing with these kinds of remarks had been a hallmark of Shannon's life up until this point. Shannon had always thirsted for adventure, even before her arms and legs had mustered the strength to crawl, but the world seemed intent on keeping her in her "place."

Shannon Wells—later to become Shannon Lucid—hadn't been born in Oklahoma but, rather, in Shanghai, China, where her parents were two expats who'd met and fallen in love there. Her mother, Myrtle, had mostly grown up in China as the daughter of a missionary doctor who'd moved his family across the world to treat patients with leprosy and tuberculosis. The adult Myrtle happened to attend a Christmas party in Shanghai in 1940. There she met a Baptist missionary named Joseph Oscar Wells, who'd just come to the country out of college.

That night, Joseph, who went by Oscar, made a bold prediction to Myrtle. "We're going to end up getting married," he told her. Oscar officially proposed on Valentine's Day, and they were married on June 1.

Less than two years into the marriage, Shannon was born on January 14, 1943. And just six weeks after she'd entered the world, she and her family were captured by the Japanese army and interned in a concentration camp at Shanghai's Chapei Civilian Assembly Center during World War II. That time is lost to Shannon, whose mind was too new to retain memories from the internment. The family wound up living in the camp for roughly a year, and then eventually boarded a Japanese vessel that took them to India—the first leg of a trip to return them to the United States. Myrtle only had one diaper for Shannon the entire voyage.

When they reached India, the family boarded the MS *Gripsholm*, a large Swedish ocean liner in service to the US State Department that was used for years to conduct prisoner exchanges. The boat carried a mix of American, Canadian, Japanese, and German POWs, all headed to various exchange points throughout the world. While on the vessel, the Wells family sailed around the globe twice as they made their way back to the US. They stopped at a port in Johannesburg, South Africa, along the way, where Shannon received her first pair of shoes. As a very young child, Shannon knew life as constant motion. She figured that was just how most families lived.

The Wells family eventually arrived in New York Harbor and waited out the remainder of the war in the United States, but they returned to China once the fighting stopped, when Shannon's inquisitive mind was newly stimulated.

It would ultimately be travel by air that would capture Shannon's fascination. When she was five, her family briefly moved to the mountainous village of Kuling, to escape the oppressive summer heat of Shanghai that might have proven fatal to her pneumonia-prone sister. For the first leg of the trip west, they piled into a DC-3 leftover from the war. Shannon's mother, brother, and sister all struggled to hold back their lunch in the turbulent air, but not Shannon. She peered out the window in awe, fascinated by the wispy clouds swirling around the mountain peaks. As the plane came in for landing on a tiny gravel runway, she spotted a little spec on the ground. The speck grew bigger, and Shannon realized it was a person, and that soon they'd be on the ground standing next to him.

"I saw this figure, this person standing down there with a red scarf around his neck and I thought it was the most amazing thing in the world—that a human being, the pilot, will be able to get the airplane down there," Shannon recalled. At that moment, Shannon made an easy decision: she was going to learn how to fly planes herself someday.

The Wells's second stay in China didn't last very long. While Shannon was in kindergarten, the Communist Revolution occurred and the family was expelled from the country. They settled for good in Bethany, Oklahoma, but Shannon couldn't quite accept their sedentary lifestyle. One day in the family's kitchen, she asked her mother, "Why don't we ever go anywhere? Why are we just sitting here?"

"But it's wonderful!" Myrtle replied, much happier with her new stationary life.

"No, it's not! I need to get moving!" said Shannon.

As it turned out, she'd find a new way to travel without ever having to leave the house. In grade school, she picked up her first science fiction novel and escaped into the distant depths of the cosmos. She was hooked. She consumed tale after tale of spacefaring civilizations and astronaut heroes exploring the universe, picking up a new book as soon as the old one was finished. Not long after her love of sci-fi began, she learned of Robert Goddard, one of the earliest pioneers of rocketry (and the namesake of NASA's Goddard

Space Flight Center in Maryland), who'd conducted test flights of prototype rockets in New Mexico. Soon, Shannon was building rockets of her own—or, more accurately, *models* of them—and she'd train in the attic inside her own cardboard spaceship, one she was determined to fly all the way to Mars.

It became an all-consuming passion. When her uncle visited from Michigan one day, she regaled him for hours on the topic of rockets and why the US should have a space program. "Don't you talk about anything else?" he asked.

Shannon never let go of her love of space exploration. Desperate to go into space herself, she read a newspaper article at the age of twelve that suggested the Soviet Union would be sending humans to space soon. She'd found a way off the planet! she thought. Delighted, she cut the article from its main page and showed it to her mother, coming to an important conclusion.

"I'm going to have to become a Communist to go to space!" Shannon observed.

The declaration didn't go over well in the conservative Wells household.

It wouldn't be long before Shannon got her wish of a US-led space program. One day, she saw seven men grace the cover of *Life*. Within the magazine's pages, they were hailed as heroes and trailblazers. The men were part of NASA's pioneering Project Mercury, a program that would pack the first Americans into rockets to screech beyond Earth's atmosphere and travel into the void of space. The group would be dubbed the Mercury Seven, and they'd go on to achieve many of the United States' biggest milestones in human spaceflight.

When Shannon saw the cover, she instantly noticed a trend.

There was a John, an Alan, a Gus. Only white men had been picked to be part of this elite spacefaring group. Her heart sank. *I'm totally excluded.*

Not content to accept these unfair circumstances, she penned a letter to the magazine, asking its editors to explain why America was only sending men into space. *Are all Americans included?* she wanted to know. Surprisingly, she received a one-sentence reply from some unknown editor.

"Someday, maybe females can go into space too," Shannon recalled the letter saying.

"Someday" seemed to be taking a while, though, both in the field of space exploration and in various scientific disciplines on Earth. When Shannon graduated college, she couldn't find any chemistry-oriented employers to give her a chance, just as her professor had predicted. The first job she could manage to get right out of school was working the midnight shift in a retirement home—a far cry from the field of chemistry. "You just had to take up the crumbs that were left, that no one else wanted to do," Shannon would later say.

Eventually, an opportunity opened up. A lab employee at the Oklahoma Medical Research Foundation walked out on the job suddenly, and the institute was desperate to fill the position. Conveniently, there was Shannon, ready and willing to fill in. The nonprofit hired her as a technician in the cancer research program, and Shannon got her first taste of real lab work for a couple of years. While she loved working there, she noted that "discrimination was alive and real," and there was essentially no way she could advance.

When a key grant fell through and Shannon was told her job would be ending in two weeks, she had to act fast. She had payments due on a very important vehicle she owned: a Piper Clipper airplane.

Once she'd graduated high school, Shannon had followed through on the pledge she'd made back in China, when she saw that tiny man in the bright red scarf. She took flying lessons, got her pilot's license, and eventually saved enough money to buy herself the tiny Clipper. She and her father would fly in it together to church meetings. But to keep the plane in the air, Shannon had to find another source of income—quick.

She sent out résumé after résumé, yearning for a response. Since Shannon wasn't a very common name back then, she'd sometimes get letters addressed to Mr. Shannon Wells. She soon learned those would be the nicer responses she'd get. One time a company asked her to send a picture of herself. When she did, she got a letter by express mail telling her explicitly that the company had absolutely no jobs available. None.

"People weren't hiring women back in those days," Shannon recalled. "And when I tried to get on with the federal government and their science labs, they wouldn't even look at a female."

Finally, Shannon went to an employment agency in Oklahoma City and proclaimed that she was desperate. Did they have anything that'd be right for her? The agency told her that a position had recently opened at Kerr-McGee, an oil company based in the city. An employee at the company was about to leave for six months to undergo training with the National Guard, and Kerr-McGee needed someone to fill in while he was away. It wouldn't be a permanent position, though. The man's job would be waiting for him when he got back. But Shannon could take his spot in the meantime.

Shannon happily accepted, even though she recognized an unfair situation. "I was coming in as a college graduate with some master's courses working in a job for somebody that I think was a college dropout."

Still, she was grateful to be employed and working in a field that actually put her chemistry skills to use. While filling in at the company, she worked under multiple men, one of whom was named Michael Lucid. He'd originally thought the company was wrong to hire Shannon for just six months, since she was so overqualified for the position. But since he wasn't the boss, his opinion had been over-ruled. He didn't say much to Shannon while they worked together.

Six months sped by, and just before Shannon's employment period was up, one of her bosses approached her and asked what her plans were. At the time, Shannon had been sucked back into another frantic job search, a pastime that was becoming routine, and she told him that she was in serious need of employment. He said that Kerr-McGee was open to hiring her full-time and quoted a measly starting salary. Shannon responded that she knew her coworker David, who'd been working in the same lab, was making much more than that, though he had less education.

Her boss just stared at her. "Shannon, you're a female. There's no way we're going to pay you the same that we're paying anybody else." She didn't have much of a choice. A quick look at the calendar showed she'd be out of a job in two weeks. She agreed, once again, to another unfair employment situation.

After coming on in a more permanent position, Shannon got an unexpected call in February 1967. On the line was Mike, her former boss, and he wanted to know if she was interested in going to the boat show with him that weekend. The request came as a bit of a shock. Shannon had absolutely no idea that he was ever interested in her. She also had plans to take her plane out that same weekend.

But bad weather prevented her from flying, so she went to the boat show with Mike instead.

The pair started dating, and it didn't take long before Mike suggested they get married. When he first brought it up, Shannon told him there was no way that was going to happen. "I plan to be a person—not someone's wife," she said. Shannon had been burned too many times—throughout her childhood and her career—and she knew that married women were expected to stay home and do . . . something. "I never could figure out what people did at home," said Shannon.

Mike, who'd grown up in the same restrictive decade of the 1950s, told Shannon that there was absolutely no reason she couldn't work as much as she wanted while she was married. He'd fallen for the woman he'd met at Kerr-McGee, and he didn't want to marry a completely different person.

Shannon then confessed to Mike a secret desire, one she'd harbored since she was a little girl. She wanted to work for NASA someday—as an astronaut. What did he think about that?

The proclamation didn't sway him. "Absolutely, no problem."

Michael and Shannon got married at her home in December 1967. And almost instantly the new family started to grow. They conceived their first daughter on their honeymoon in Hawaii, whom they would name Kawai Dawn, after the island they'd visited. The newlyweds lived in a state of pure bliss those first few weeks post-nuptials.

Then Shannon informed Kerr-McGee of her pregnancy, and her bosses immediately fired her.

The joys of marriage were quickly overshadowed by the familiar sting of unemployment—at least for Shannon. Kerr-McGee didn't see any reason to fire Mike. He continued to go to work to make money for the family, and for weeks Shannon would stand at the door crying as he left each morning. Finally, Mike declared that this routine had to end. "We can't live like this," he told her. So he made a suggestion: Why didn't Shannon go to graduate school and get an advanced degree? In such an unfair world as the one they inhabited in the 1960s, Shannon clearly needed to accumulate as many credentials as she could to find long-term employment.

So Shannon returned to the University of Oklahoma to get both her master's and PhD in biochemistry. During that time, she and Mike had their second daughter. When Shandara was born, Shannon was back at school less than a week later to take her finals. "I couldn't put off the test," Shannon explained.

It took four years for Shannon to achieve the pinnacle of academic rank, and even after she'd graduated with her PhD, finding employment *still* turned out to be a struggle. She suffered again through months of job searches, sending out round after round of government applications. But she ultimately landed at a familiar place. The Oklahoma Medical Research Foundation took her on again in 1974, this time as a research assistant to study how various chemicals caused cancer in cells. She stayed for the next few years, finally content in a full-time position.

Then one day in July 1976 Shannon was working in the lab, reading the foundation's science magazine, when she spotted a short article toward the back. The blurb mentioned that NASA was recruiting a new round of astronauts for the agency's new Space Shuttle. And this time, the agency wanted women to apply.

ANNA LEE (SOON to become Anna Fisher) slumped down in a chair at Harbor General Hospital in Los Angeles, completely exhausted. She was in the midst of another twelve-hour day at her surgical internship, a gauntlet that had become pretty standard for her after graduating from UCLA's medical school. The entire last year had been like this, filled with back-to-back shifts on her feet and calls to perform last-minute surgeries at any wild hour of the night. She knew these were the dues she had to pay to become a surgical resident, but she'd started to wonder if this was really the kind of life she wanted.

As she took a second to relax, she heard a familiar voice over the hospital's PA system. It was that of her fiancé Bill Fisher. He was paging Anna to call him back. *I wonder what this is about?* she thought.

Anna wrenched herself from her chair and made her way to a nearby phone. She looked around Harbor General as she walked, the place she'd spent almost all her time this past year. In truth, her relationship with the hospital went back even further. Anna had volunteered at Harbor General while she was at San Pedro High School in the late 1960s, along with her friend Karen Hayes. The two had been candy stripers at the hospital. The name was a nod to the pink-and-white pinafore uniforms that women wore during their volunteer work. It had been a particularly daunting introduction to the world of medicine, as the hospital's wards were filled with severely ill cancer patients, and their pain weighed on Anna. So she and Karen opted to sequester themselves by developing X-ray photographs in the hospital's darkroom, a pretty important gig for two high school students.

It was during their volunteer work that Anna had admitted something huge to Karen. One day the two friends stood next to each other in the dark, flipping film and plunging the photographs into their processing chemicals. Perhaps it was the absence of light that made Anna feel comfortable sharing her deepest desire. The darkness obscured her friend's face, masking whatever facial expression she might have in response. But in that moment she confessed to Karen something she'd never told anyone. "I'd really like to be an astronaut someday," Anna admitted, as the two talked about their futures beyond high school.

Karen was shocked, since her friend had never betrayed a hint of that aspiration. But the truth was that Anna had been thinking about it since junior high. Specifically, she'd been thinking about it since May 5, 1961.

That day, a man by the name of Alan Shepard donned a silvery space suit and a white helmet in Cape Canaveral, Florida. Just before four in the morning, he boarded a white van that dropped him off in front of a large black-and-white rocket, hissing and venting gas while it waited on a red metallic launchpad. Emblazoned on the side of the vehicle were the words UNITED STATES in bright red letters. With a portable air-conditioning unit in his right hand, Alan stepped out of the van and peered up at the rocket, a Redstone, one last time.

He then climbed into a capsule perched on top of the missile and, hours later, took off into the sky. The rocket propelled his Mercury capsule, dubbed *Freedom 7*, to a height of 116 statute miles above Earth's surface, before it dropped back down and splashed underneath a parachute in the Atlantic Ocean. From takeoff to splashdown, the entire flight took just fifteen minutes. But in that short interval, Al became the first American to reach space. He hadn't been the first man *ever*, though. That title went to Yuri Gagarin, a Soviet pilot who'd successfully orbited Earth less than a month earlier.

Anna, nearly twelve years old at the time, listened to the entirety of the flight on a transistor radio outside her school at Fort Campbell, an army base straddling the Kentucky-Tennessee border. She and her classmates sat on their playground's dewy grass, all packed in tight around the radio, hoping to puzzle out the scratchy audio from the announcers. Anna had been worried she might miss the launch entirely that morning. The flight was originally supposed to take off at 7:20 a.m. EST, when she had to get ready for school. But thick clouds above the launch site pushed the launch back a couple of hours, putting liftoff right in the middle of her first-period PE class. Luckily for Anna, her teacher paused the day's fitness tests so that the kids could experience history. (It wasn't so lucky for Alan Shepard, who wound up relieving himself in his space suit after four hours of waiting for the launch to get off the ground.)

Those fifteen minutes of spaceflight forever imprinted themselves on Anna's heart. "As he launched and I listened to him, I decided at that moment that, if I ever had a chance, that's what I wanted to do," she said. For her, the job was the perfect combination of everything she enjoyed: it involved science, it was challenging, and it involved exploring the unknown.

It was a moment of clarity in a life defined until that point by motion. Her father, Riley Tingle, was a sergeant in the US Army and would uproot the family each time he got reassigned to a new military base. Such a life had its benefits. For one, Anna wouldn't have been born if her father hadn't traveled the globe. His career led him to Berlin after World War II, where he met Elfriede, a German-born woman working for the US military—a job she secured thanks to her extensive knowledge of English. They married in April 1949 and came back to the United States together, where they had Anna. As soon as she entered the world, she found herself in a new hometown every couple of years. A shy little girl who enjoyed doing her math homework each night, she struggled to make friends and to shape her own identity.

It wasn't until that moment outside her school in Fort Campbell that Anna had true resolve on the kind of path she wanted to take—though she immediately thought her dream would never be possible for her. She, too, had noticed that only men had been selected to be astronauts thus far. Plus, all the astronauts had to have jet piloting experience, something you could only get by joining the US military. And women were prohibited from flying jets for the armed forces.

When Anna turned thirteen, she finally got a reprieve from the transient lifestyle when her family settled in San Pedro, California, a town populated by immigrant families from across the globe. Anna would later be grateful that she spent the remainder of her childhood in one place. The junior high school she attended in San Pedro was her thirteenth school, which meant she'd averaged about one new school every year of her life. She welcomed a break from learning the names of a whole new crop of classmates every couple of semesters.

By the time Anna had moved past her candy striper phase and gone off to college in 1967, her dream of flying to space had been pushed deep into the back of her consciousness. And that meant thinking practically about her future. She was accepted at UCLA, the one school she applied to, having done so well on her SATs that the school awarded her a merit scholarship during a ceremony in front of the entire student body. She was the first in her family to go to college, and as such, a bit of an anomaly for women in those years. Her teachers didn't exactly encourage women to seek college degrees, and most women in Anna's school didn't aspire to more schoolwork or full-time employment. They'd been told most of their lives they wouldn't need such things when they grew older.

Anna initially chose math as her route, but she started to wonder what she would actually *do* with a degree in advanced mathematics. She ultimately decided she didn't want to be a math professor or theoretical mathematician, which meant she should probably consider another path. In the meantime, she'd taken a few chemistry classes, which had really intrigued her. So she made the leap to a chemistry degree, figuring that she could do more with a science education.

Her time at UCLA had been illuminating and intense. She'd had a brief whirlwind romance, both marrying and then later separating from a man who turned out not to be the right partner for her. And as her undergraduate years came to an end, Anna decided she'd apply to medical school at UCLA, instead of doing research with her chemistry degree. While she loved the idea of a career as a doctor, there was a hidden motive for this move. Still dreaming of going to space one day, she figured that eventually there'd be space stations orbiting the Earth—and those stations would need doctors. Just like Rhea had imagined, Anna thought perhaps she could be the one to live at those stations, attending to the sick astronauts. It was a unique gamble that she kept to herself during the admissions interview.

Even though she kept her astronaut ambitions to herself, UCLA rejected her application initially, putting her on the wait list. For Anna, it was a blow, but she decided to make the most of that time. She pursued her master's in chemistry while working as a teaching assistant. At the time, it seemed like a failure to wait, but the degree would come in handy in the years to come.

When she did get into medical school, she opted to attend Harbor General for her internship, a place that was familiar to her. And it was there that she met Bill, another surgical intern who was a year ahead of her. They met in the hospital's cafeteria when Anna overheard a group laughing at Bill, who was loudly telling a few blush-inducing jokes. He looked more like a movie doctor than a real one, with a chiseled jawline and a full gleaming smile. The two also had quite a few things in common. Of course, they shared a passion for medicine, but it also turned out that both harbored a deep desire to become an astronaut, something they learned on their first date. The two would talk all the time about how if the opportunity ever came up, they'd jump at the chance to apply to NASA. The couple quickly became smitten with each other, and Bill proposed less than a year after they met.

Anna eventually found a phone to call her fiancé in Harbor General that day after getting his page. Bill's exuberance seemed to leap out of the phone's speaker. He explained that he'd just had lunch with a mutual friend, Dr. Mark Mecikalski, who disclosed a major piece of information. An avid fan of NASA, Mark had learned that the agency was about to recruit a new round of astronauts, with women encouraged to apply. And since Anna and Bill were always talking about working for NASA one day, he figured they should know about the selection.

The news filled her with newfound energy. This was the moment Anna had been waiting for ever since she was a little girl, and here she was with the right credentials to give it a shot.

There was only one problem, Bill told her.

"We have three weeks to apply before the deadline," he said.

CHAPTER 2

Far from Home

Kathy Sullivan could taste the saltiness of the chilly ocean waves as they crashed against the side of the boat carrying her out to sea. She stood on the deck of the CSS *Hudson*, Dalhousie University's white research vessel, which was currently speeding into the wide-open waters of the northern Atlantic. The crew's focus of interest was actually located below the frigid waves. Kathy and her expedition team were headed to the Grand Banks, a submerged chain of plateaus off the coast of Newfoundland in Canada. There, long-dormant volcanoes dotted the edges of the underwater terrain, and Kathy and her crew were going to get a taste of these ancient peaks.

Once the *Hudson* found itself floating above one of the sunken summits, Kathy hoisted a cable overboard, dangling it under the surface of the water. The goal was to gently scrape the tops of these mountains, capturing a small sample of material to bring back home. Kathy and her team were eager to know what kinds of chemicals had been deposited in these old volcanic rocks; they might provide some insight into how the seafloor of the Atlantic Ocean first formed.

For Kathy, nothing could compare to being out at sea like this. She loved the thrill of a successful expedition. To her it felt like orchestrating a beautiful song, the symphonic culmination of months of planning and coordination. But most of all, this was the kind of adventure she had always dreamed of as a girl—standing out here on a boat surrounded by nothing but crystal-blue ocean, the northern wind whipping through her hair.

From as early as she could remember, Kathy had wanted to explore the unknown. But it wasn't the vast realm of space that first intrigued her; it was the planet she lived on. Maps had been her gateway to this emerging passion. She devoured the intricate landscapes and riverbeds speckled across the European continent—and the vast mountain ranges and distant oceans of Asia. By age seven, Kathy had created her first map. She soon learned the intricate language of cartography, decoding the legends and cardinal directions of the maps she'd find in her *National Geographics*.

It was just her nature to figure things out and unravel the mysteries of the world around her. When she was in kindergarten, she'd seen a television ad for a plastic toy pom-pom gun, one that simulated the recoil of a real gun turret located on the deck of a battleship. She *had* to have it. Not because she was a burgeoning gun aficionado. She wanted to get her hands on it so she could see how it worked—how its levers actually moved in that knee-jerk way.

Her zest for engineering likely trickled down from her father, Donald, who also introduced her to the world of aerospace. When Kathy was six, her father accepted a job as an aerospace engineer at the Marquardt Corporation in Van Nuys, California. The company was involved in the development of the ramjet engine, and suddenly planes were a staple of family life. Donald would bring home schematics and blueprints that Kathy would trace with her fingers. Marquardt's headquarters happened to sit right next to the Van Nuys Airport, and Donald soon joined the company's flying club, taking Cessnas and Pipers into the air to join buddies on fishing trips. Kathy's brother, Grant, also caught the flying bug and would often accompany their father in the air.

Despite a burgeoning love for space that was fed by following America's developing space program, Kathy felt particularly drawn toward Earth-bound travel. Such travels would have to wait, though, because she'd be needed at home during her formative years. When she was twelve, her maternal grandmother and only living grandparent died of cancer, causing Kathy's grief-stricken mother, Barbara, to turn to alcohol. Kathy watched the mom she once knew transform into someone unfamiliar, a crushed woman plagued by grief. And Kathy found herself stepping up to help stabilize the family. She'd take on household cleaning tasks and play mediator during tense fights. She and Grant would even disable the car when their dad was at work, in hopes of preventing their mother from getting behind the wheel after drinking.

Kathy yearned for a time machine so that she could travel back and retrieve the mother she once knew. Eventually, thankfully, her mom would seek therapy and emerge from her depression, though life was never quite the same again in the Sullivan household.

As Kathy approached the end of high school, she set a concrete goal for herself: She was going to live abroad. And to do that, she figured that foreign languages would be her passport, granting her access to countries that would otherwise be off-limits. That language-acquisition journey didn't last long, though. During freshman year at the University of California, Santa Cruz, an advisor insisted that Kathy had to take at least three science classes. It was *not* something Kathy wanted to do. Her advisor held firm, though, and he suggested a couple classes she might enjoy, such as oceanography and geology.

The lessons introduced Kathy to a world she hadn't known existed. One of her professors would take students out to the tide pools along the California coast, looking for strange ocean critters like sea squirts. Kathy found the rhythms of the various tides and currents fascinating and was equally captivated by the intricate rocky structures stretching across the seafloor and the ways that the land and the sea intertwined. She felt unexpected excitement when she learned that the people who studied the oceans got to explore the world—it was required in order to do their jobs.

She eventually mustered the courage to ask one of her oceanography teachers what the job of an oceanographer entailed. He took her under his wing, patiently answering her questions and suggesting more classes she could take to indulge her curiosity. A few courses later, Kathy made the switch from languages to Earth sciences. The pivot still allowed her to live abroad. In her junior year, she studied at the University of Bergen in Norway.

It was an exciting time to be a student of the oceans. Just a few years before Kathy had entered college, geologists had begun accepting the concept of plate tectonics and continental drift—the constant movement of gargantuan landmasses like so many puzzle pieces fitting together. It was a concept that opened up a ton of research opportunities for young oceanographers. And Kathy got to ride that wave in Norway, and then later at Dalhousie University in Nova Scotia, as she worked toward her PhD.

She'd spent three and a half years at Dalhousie so far, venturing out to sea on the *Hudson* numerous times. But as she stood on the deck of the boat, scraping up rocks from the seafloor, she knew what she really wanted was to dive *below* the surface of the ocean. On a previous expedition, she'd been on a vessel that had held the *Alvin* submersible, a mini-submarine capable of transporting humans down to extreme ocean depths to properly map the seafloor. As soon as she saw two researchers descend inside that vessel, she knew she wanted to one day pilot *Alvin* herself. Her goal was to travel as deep as she could below the Earth's surface.

Still, Kathy was having the adventure of a lifetime on the Atlantic's surface near Nova Scotia, and she brought tales of her expeditions home with her when she returned to California over Christmas in 1976. She confided to her brother her desire to dive in *Alvin* someday. But as he listened to her talk about that and other goals, he suggested another career path, one that would take her, quite literally, in the opposite direction: Had she considered applying to NASA's new astronaut selection round? They were trying to recruit women and minorities this time, he told her.

Kathy hadn't heard anything about that. She'd been living mostly in Canada, and news from the States didn't percolate northward very often or fast. Since Grant was the family's aviation buff, he'd already applied. But he pestered Kathy throughout her break to join him in the application process. "How many twenty-six-year-old female PhDs can there be in the world?" he asked her.

At first, she thought it was an absurd idea. She kept thinking about how hard it was to study the ocean floor from just the *surface* of the Earth. Putting hundreds of miles between her and the ground would only make things worse. So she brushed off Grant's pleas, returning to Nova Scotia with no plans to apply to some silly astronaut program.

Then a few weeks later, Kathy was thumbing through a science magazine when she saw an advertisement about the NASA selection process. The article laid out some key facts she hadn't been aware of: the astronauts selected in this round would be flying on a brand-new type of vehicle known as the Space Shuttle, which would be geared toward launching satellites and conducting science experiments. And with science at the forefront of these missions, NASA would be looking for more than just pilots. The agency needed people to fill the new role of "mission specialist." These astronauts would work primarily on science-related projects and wouldn't need any flying experience.

At that moment, Kathy realized that her voyages at sea might perfectly translate to voyages in orbit . . .

———————

SALLY RIDE PEERED up at a volleyball coming over the net straight toward her. Moving quickly, she put her hands up, forming a loose triangle with her fingers, just as the ball made contact with her palms. The volleyball bounced off her hands and flew leisurely upward, giving her teammate Bill enough time to run up and smack the sphere hard over the net. It was a perfect set and spike for the duo, a combo they'd perfected during dozens of pickup volleyball games on the Stanford campus.

These games had become a near-daily occurrence for Sally and her group of friends. She'd become part of a tight foursome—one made up of two men, Bill Colson and Richard Teets, and her roommate, Molly Tyson. The group would get together practically every day to play volleyball, often followed by dinner at Sally and Molly's tiny two-bedroom house on the Stanford campus.

Life in grad school was good for Sally, and she was happy to be living in her home state once again. It was also a life that could have easily never happened. There was a time when it seemed like her future lay on the tennis court—not in academia.

Standing on a burnt-orange clay court in Spain, Sally had picked up her first racket at age ten. She grew up a California girl, through and through, but in 1960, her parents had sold their house in Van Nuys and taken their two girls on a year-long road trip through Europe, relying on a Borgward station wagon for the journey (which they'd nickname Borgy). It was an enlightening sabbatical, filled with new cultures and cuisine. But that day in Spain, she lit up as she learned how to serve and close out a set.

When the Ride family returned to California in 1961, to Encino in the San Fernando Valley, Sally threw herself into tennis. She took to the game with ease, but her knack for serving and volleying wasn't too surprising. She'd always been a skilled athlete and sports buff. It was a love she inherited from her father, Dale. An aficionado of any game with a ball, he always had some kind of sports event on the TV, and he loved to play with Sally and her sister, Karen, in the backyard. As a political science professor at Santa Monica College, he often helped basketball and football players transfer from there to UCLA. Those connections got him and his daughters into Bruins practices, boosting Sally's love for any kind of game.

Truthfully, it was baseball that ensnared Sally from the beginning. As a California girl, she revered the Los Angeles Dodgers above all. She amassed a large collection of the team's baseball cards and, when she was just five years old, Dale taught Sally how to read and decode the team's box scores in the newspaper each morning—perhaps her first introduction to the intricacies of math.

Surrounded by sports, Sally was virtually destined to become a jock. Tennis brought Sally many things, like dexterity and good hand-eye coordination, but it also brought her a core group of friends, all sharing a love of the game. During one tournament in Redlands, California, she matched against a twelve-year-old girl by the name of Tam O'Shaughnessy. The day they met on the court, they instantly clicked. Tam's mother and Sally's father watched frustrated as the two girls spent more time gabbing with each other than trying to close out a set. "Sally and I, it just wasn't competitive," recalled Tam. "We were more interested in getting to know each other and chatting and drinking water [while] changing sides than actually playing the match. And our parents were like, 'What are they doing?'"

Sally would often trek down to San Diego to meet up with some of her tennis pals, Tam included, or her friends would come to Los Angeles to see the Ride family. Together they'd practice their drop shots during the day and at night just hang out, as young girls do. After a long day of outdoor play, the girls would usually listen to records to wind down, triggering impromptu dance parties in dens and living rooms. Sally would never be the first to get up. A classic introvert, she usually sat in the corner, not wanting to draw attention to herself. Tam would spy Sally huddled alone and tug on her hand, pulling her into the center to include her in the group dance. It wouldn't take long for Sally to start singing along with all the rest.

This tendency to hang back would translate to her tennis game too, though it wasn't about a lack of courage. Sometimes Sally just wanted to be a little lazy. She'd often blow off her commitments to sprawl in front of a television screen, with her brain turned off—a pastime that would irk Dale. He'd encourage her to get up to practice more, but Sally wouldn't budge. She valued her downtime as much as she valued achieving the perfect ace.

But despite these periodic lapses in motivation, she still became a powerful player, inspired by the tennis greats of the era like Billie Jean King, Pancho Gonzalez, Maria Bueno, and Rod Laver. Her tennis skills would eventually secure her a place on the team at the ritzy Westlake School for Girls, attended by children of wealthy business magnates and celebrities from the Los Angeles area. Since the school was far out near Beverly Hills and Bel-Air, she joined a carpool with a classmate by the name of Sue Okie, a tall girl who couldn't get enough of Sally's flashy white smile. The two became inseparable as they cruised down the California highways together to get to school.

Sally was a great student—when she wanted to be. But just as she'd lightly blow off her father's calls to practice more, she'd also procrastinate in doing her homework, sarcastically priding herself on being an underachiever. "She was certainly able to get straight As if she wanted to, but she kind of made it a point of pride to not kill herself working all the time—you know, to not necessarily be completely prepared for tests," Sue said. If Sally didn't like something, she wouldn't throw herself into it. This was especially the case when it came to her English and history classes, where she dreaded being called on by her teachers.

Science classes—they were a different story.

Sally had gotten her first taste of science and space when she was young, after her parents had bought her a small Bushnell telescope. She'd peer through its glass at night, pinpointing the constellations she could find. Her favorite had been Orion's belt. With just three stars, it was an easy one to spot. And Sally would pass along her fascination with the sky to her friends. "We slept out on the lawn one night, and we're looking at the stars and we're talking about, you know, the enormousness of the universe and how many years it takes for light to get here," said Okie. "And her imagination was really captured by all of that."

What really sent Sally's passion for science into overdrive was a woman named Dr. Elizabeth Mommaerts, a physiology teacher who taught Sally and Sue in eleventh grade. Dr. Mommaerts had been a professor at UCLA, a rare woman with a PhD at the time. Plus, she was just plain cool. She was Hungarian and cosmopolitan, bringing a college-level understanding of science to her high school students. Sally was enamored with the teacher and the way she taught, while Dr. Mommaerts saw the vast potential bursting within Sally. The two would challenge each other during class, especially if Sally came in unprepared (which did happen on occasion).

Dr. Mommaerts would host dinner parties for her best students, which Sally would attend with pleasure. Sometimes, Sally would bring her teacher a brain teaser, which Dr. Mommaerts would quickly solve, offering Sally some sort of riddle in return. Sally very much cared about how Dr. Mommaerts perceived her, and though she wasn't a particular fan of physiology, she took from her a deep love of science. Dr. Mommaerts encouraged her to pursue a career in research, whatever kind it might be.

All these paths ultimately led to Sally choosing physics as her major when she headed off to college in 1968. Thanks to tennis, she secured a full scholarship to Swarthmore College in Pennsylvania. But it didn't take long for her to realize Swarthmore wasn't the right place for her. She'd loved the school, and the classes really nourished her burgeoning love for physics. But she grew homesick for California, where the weather was much more cooperative. And after dominating the tennis tournament circuit in the Northeast, she'd had a small epiphany. Could she *actually* make a go of professional tennis before it was too late? She'd been on an insane winning streak, dominating match after match and securing the Eastern Intercollegiate Women's Singles Championship.

So she dropped out of Swarthmore after three semesters and went home to see if such a life was possible. No more ditching practice when she didn't feel like it. It was time to buckle down. Once back in the Los Angeles area, she joined the UCLA tennis team after transferring to the university and got to work. But the quest to go pro was just as short-lived. It had only been a few months of backbreaking tennis before she got a wake-up call: her heart just wasn't in it. She found she couldn't fully commit to the packed days of practice it would take to become a master. And over time, she realized that her love of science and education was much more important than becoming a tennis star. Up until then, Sally had been walking two paths—one leading to athletics and the other toward science—and she'd made the decision to focus on the route that led to academia.

During her time at UCLA she'd taken a course in Shakespeare and another in elementary quantum mechanics. Liking them both—yes, even the English class—she decided to transfer to Stanford University, where she double majored in physics and English. Her tennis-playing days weren't over, either. The dean of admissions, who happened to be the same one who'd accepted her into Swarthmore, happily accepted her to play on the Stanford tennis team.

Sally truly loved Stanford, with its arched sandstone buildings nestled among the redwoods and rolling hills of Northern California—so much so that she stayed on for graduate school, specializing in physics. She was undeterred by the fact that she was one of just a handful of women in the entire physics department. When a professor told Sally his class had a 60 percent dropout rate, she fought her fear and persevered until the semester's end, watching her fellow students suddenly disappear one by one.

After a few months, she began narrowing her focus to the study of astrophysics—how celestial objects like distant stars and planets interact and form throughout the universe. It was a natural extension of her childhood hobby, looking up at the night sky to trace the twinkling stars and constellations above her. She was putting everything she had into her studies—no longer a self-proclaimed underachiever—and loving her research in a new field of study that centered on free-electron lasers, which involves blasting beams of tiny particles called electrons through a magnetic field, triggering bursts of radiation.

All that *and* she had her volleyball team.

After another game one day, Sally was home at the two-bedroom house she shared with Molly when there was a knock at the door. She opened it to find Bill, her friend and designated spiker, standing there. He asked to come in and talk to her. The two sat on the couch, and an obviously nervous Bill finally blurted out what he'd come to say. He had feelings for her. The two friends had been spending an awful lot of time together, and Bill had started to think of Sally as more than just a friend.

Sally was quiet for a minute. "This is really difficult," she said. After another pause, she revealed why. "I'm in love with Molly."

It had been the first time Sally had admitted that to anyone. Until that point, her romantic relationship with Molly had been conducted strictly behind the scenes. They'd begun as friends, as occurs with most relationships. The two women had known each other as kids through the junior tennis circuit. And when Sally had come to Stanford, she'd found Molly after some unauthorized sleuthing. She'd gotten a job at the registrar's office to help pay for college, and had used her coveted position to look for names of students she recognized who were enrolled in the school (a forbidden pastime). When she spotted Molly Tyson on the list, she recognized the name instantly and wrote down her address.

Molly was shocked when she got a knock on the door and found her old friend Sally Ride on her doorstep. They hadn't even really been *that* close as kids, since Sally had been a year older and ranked higher than Molly. But soon that didn't matter. They hit the court and played tennis and almost instantly became an inseparable pair. When the two joined the Stanford tennis team, Sally insisted on playing doubles with Molly, even though Sally was ranked first and Molly sixth. They did everything together, from going on weekend trips to Tulsa to composing a long silly children's book on butcher paper. And since Molly was also an English major, they'd quote Shakespeare to each other as inside jokes.

It had begun as a friendship but evolved into something more. Just after transferring to Stanford in 1970, Sally suffered her first big heartbreak, when her first real boyfriend broke up with her in a letter while on a research trip to Moscow. Molly had been there to comfort her, and that comfort grew into a romance. For both, it was their first time being with another woman. And as such, they kept their newfound love under wraps. Attitudes toward gay relationships may have been changing in the 1970s, but there were still a fair number of people who viewed queer couples as deviants. The two never talked about it openly, but it was understood that they'd keep their love a secret. Plus, uncertainty was a factor. Both Molly and Sally had only dated men until then, and neither was about to proclaim they were openly gay after this one relationship.

Their discretion was so complete that Bill had never suspected anything. He took the news in stride. He didn't judge Sally—rather, he was mostly bummed that a relationship with her was no longer in the cards. The next day back at the house, Molly found him and asked him about his talk with Sally, confirming what Bill had been told. *Well, this is all over*, Bill thought.

But what Bill didn't know at the time was that Molly and Sally were nearing the end of their relationship. By that point, they'd been seeing each other in private for close to five years, and the two had shared incredible moments together. While working as camp counselors at a tennis camp in Lake Tahoe, Sally and Molly got to meet Billie Jean King when she stopped by the camp. Molly watched as Sally played in an exhibition match with King, who'd won Wimbledon four times by that point. King later complimented the Stanford student on her "good backhand," leaving Sally elated.

It had been a remarkable journey but Molly had come to feel resentful. Sally was almost godlike in some ways—a great athlete and scholar—and it grew harder to date someone with such legendary talents. Molly also wanted to be more open. She was tired of living in the dark and eager to shine a light on her true life with Sally. But Sally wasn't ready for the world to see her that way. Their dueling desires led to an inevitable split. Molly escaped to the opposite coast, moving to New York to become a sportswriter, leaving behind California and Sally, who found herself heartbroken once again.

It would take years for Sally to truly get over Molly. But she moved on with her life while at Stanford. Bill Colson had started dating other women, and when Sally spotted him out with someone else, she went to his office the next day to let him know she was jealous. That declaration reawakened Bill's interest, and the two became an item, falling back into their routine of spending all their time together.

In January 1977, Sally had a year and a half to go at Stanford, and she began thinking about what lay ahead after graduate school. She and Bill had talked about it. The couple figured that, upon graduation, they'd probably go off and become professors somewhere. But neither had fixed ideas of what mountains they wanted to climb.

Then one morning that January Sally walked into the Stanford student union to get a cup of coffee and some breakfast before class. She reached for a copy of the *Stanford Daily* as she took a bite of her meal, and her eyes lit up as she read an article on the paper's first page:

NASA TO RECRUIT WOMEN.

JUDY RESNIK SAT in the passenger seat of a sleek Triumph TR6 sports car, staring intently at the calculator in her hand. Next to her in the driver's seat sat Michael Oldak, her husband, diligently maneuvering the car on a back road in New Jersey. Michael kept his foot steady on the gas pedal, trying to maintain the speed Judy had just calculated for him. They kept their eyes peeled for the upcoming checkpoint in the ground, hoping to reach the marker at a very specific time.

The two were in the middle of a race, but they weren't trying to gain ludicrous speeds. Instead, the goal was punctuality. Judy and Michael had become avid fans of time-speed-distance (TSD) sports car rallies, a nerdy sport where drivers maintain just the right speed—not too fast, not too slow—to reach various checkpoints at predetermined times. It's ultimately a team sport, requiring both a driver and a skilled navigator who can do the math in real-time. Judy's brain made her the perfect guide. The race was just another average weekend for the couple, a way to unwind when they weren't working together at RCA.

It was also the perfect hobby to satisfy Judy's brilliant logical brain. Born on April 5, 1949, Judy learned to read and solve math problems in her first five years of life. She was so advanced, in fact, that her kindergarten teacher recommended Judy skip to first grade. The school psychologist figured she could handle it, and Judy soon became one of the youngest kids in class at Fairlawn Elementary School in Akron, Ohio.

It wasn't just schoolwork that came easy to her. Judy started taking piano lessons in grade school, studying under accomplished Ohio musicians Arthur Reginald and Pat Pace. Soon she flourished as a classical pianist, and her teachers told Judy's parents she had the chops to play piano professionally someday.

Judy excelled in everything she did, achieving a GPA of 4.2 at Harvey Firestone High School in Akron. She became the sole female member of her school's math club, a type of unbalanced group setting she'd encounter numerous times throughout the rest of her life. But her crowning achievement came when she achieved a perfect score on the math portion of her SATs. She'd go on to graduate at the top of her class.

Most boys her age would have been intimidated by the raven-haired, whip-smart Judy. But not Michael. They met during her freshman year at Carnegie Institute of Technology (called Carnegie Tech at the time). Michael's fraternity had a tradition in which the boys would visit all the new freshmen girls, handing out roses. The frat would then secretly keep a list of all the prettiest girls, of which Judy was one. Michael's roommate had decided to ask Judy out based on the list's recommendation. But when he brought her by the frat for their date, Michael decided to step in and ask Judy out himself. "She was too good for him," Michael claimed.

Michael was everything Judy could want in a boyfriend. He was a smart, driven electrical engineering major. He recognized her intelligence and yet wasn't intimidated by her brilliance. And he was Jewish, a palatable partner for her extended Jewish family. Judy's grandfather, the Rav Jacob Resnik, had been a rabbi from Kiev, Russia. He moved his family to Palestine (now Israel), where his children—including Judy's father, Marvin—learned to speak Hebrew. Eventually he and his wife, Anna, along with all six of their children, immigrated to the United States, settling in Ohio. Marvin and his brother Harold became optometrists, while many of the other kids had their own businesses. The extended Resnik family would make a point to gather every Friday for Shabbat dinner at Jacob and Anna's house in Cleveland—the grown-ups at one table, Judy and the cousins at the other.

When Judy began dating Michael, she accompanied him to a few of his electrical engineering classes and found she really enjoyed the lessons. She wound up switching her major from math to electrical engineering, making her one of three women in the whole program. The couple bonded over more than schoolwork, though. While on a date to a theme park called Kennywood outside Pittsburgh, Michael urged Judy to ride a roller coaster with him. He finally wore her down, and they spent a heart-splitting three minutes whipping up and down the steel rails. After they'd exited the ride, she turned to him and said: "Let's go again."

Overall, college had opened up a glorious new world for Judy. She found friendships through her sorority, Alpha Epsilon Phi, and eventually moved in with one of her sorority sisters, who was looking for someone who'd be willing to keep a kosher household. While such a lifestyle wasn't exactly important to Judy, she didn't mind adhering to it to help out a friend. Meanwhile, she continued to excel at everything she did, taking to her new major with ease. While Michael would stay up all night studying for his electrical engineering exams, Judy would easily be in bed by ten, confident she had all the answers for the next day's tests. Since she got top marks in all her courses, she was usually right. "She was just absolutely brilliant and extremely talented at anything she did," said Michael. Her classmates took notice, voting her runner-up homecoming queen. As a treat, she and Michael got to dance onstage with the band Chicago during the school's homecoming concert.

Carnegie had provided Judy an escape, affording her a level of freedom she'd only dreamed of while at home. But with Akron only a couple of hours away, Judy still maintained ties to her past life—one that had its share of hard memories.

From the beginning, her mother, Sarah, demanded excellence from Judy. Sarah tried to instill in Judy a sense of order and discipline by creating a very regimented life for her daughter. Judy would come home from school with her good friend Barbara Cheek to find a table filled with ingredients to bake cookies. Barbara watched in awe as Sarah hovered over Judy, dictating how to put the ingredients together, walking her through the process step-by-step. After baking class, it was time to hit the piano keys. These kinds of lessons happened all the time, with Judy's life after school very much controlled down to the minute.

"All her time was structured, completely structured, and that's what fascinated me because my life was just chaos," said Barbara Roduner, née Cheek. Such a set schedule made Judy eager to experience what other people's lives were like. "And that's one reason she wanted to come to our house—to see a different way to live."

While she learned discipline from her mother, Judy found a much more relaxed and congenial relationship with her father. When Marvin came home, noted Barbara, the energy in the room would shift. "The minute he came home from work, the kids were freed and they'd go right on his leg," said Barbara. "And he just spent an amazing amount of time with them." When company came over, Judy would immediately go to the piano to play, with Marvin accompanying her with his singing. He never skipped an opportunity to talk about how proud he was of her accomplishments.

As a result, Judy eventually drew traits from both Sarah and Marvin. She could be aloof and intense, according to Barbara, focused squarely on her studies and work. But her friends also knew her as kind and compassionate. She didn't express her feelings much, keeping her vulnerabilities to herself, but she was warm to the people she loved and available when someone close to her needed a hand. Meanwhile, her friends were there to help Judy, too, when she needed to undergo a teenage rite of passage, eventually working up the courage to sneak out of the house with Barbara and her sister, Pam, in high school.

Around that time, Judy also started dating the requisite bad boy, Len Nahmi, after meeting him at a basketball game. He was Judy's foil in a lot of ways: he didn't make great grades at his Akron high school, wasn't a hard worker, and wasn't Jewish—all to the dismay of Marvin and Sarah. They told Judy she wasn't allowed to see him, but she'd find ways to work around their command. Sometimes they'd meet up in Cleveland while Judy stayed with her cousin. Len may not have been the star that Judy was, but he did express the ambition to fly planes one day, something that intrigued Judy.

The night-and-day dynamic of her mother and father couldn't last and their marriage began to deteriorate. Around the same time, Judy's relationship with her mom also devolved. "When Judy was fourteen, all her friends were ice skating," Marvin recalled once. "She asked for a pair of ice skates and Sarah said no. So I went out and bought her a pair. Sarah burned them. Can you imagine? Next year, I bought her another pair." Adding to the friction level, Sarah still didn't approve of Len, and the young couple grew tired of the immense work it took to stay together. Len finally broke up with Judy on the phone, arguing it was a lost cause. He told her that "the whole thing was causing her nothing but anxiety and pain."

Marvin and Sarah's marriage ended in divorce when Judy was seventeen, just before she'd strike out on her own. As was usual in divorces at that time, Judy and her brother, Chuck, went to live with their mother. But before she left for school, she made a bold move for a teenager. With the help of her father, she prepared a court case, requesting to have custody switched from her mother to her father. The courts granted the request.

It was more of a symbolic gesture than anything. It was a coda to a toxic relationship, and perhaps a bid for freedom.

Judy stayed close with her father while in school, periodically writing letters to him in Hebrew and bringing home her new boyfriend during holidays and on weekends. Judy's dad remarried after she'd graduated high school, giving her two new twin stepsisters she grew to adore. Michael loved meeting the family, and he easily fit in, taking to Marvin's sense of humor.

But the visits to Judy's mother's house took on a different tone. It was clear Judy's relationship with Sarah was increasingly strained.

The emotional distance between Judy and Sarah only widened during the Carnegie years, and perhaps as a result, Judy remained as guarded about her life as ever. Though Judy's friends all loved her, everyone knew her as an intensely private person who kept to herself, making it hard for the people close to her—even Michael—to truly know her. "She was always happy and smiling and fun to be with," Michael said, with the caveat: "You really didn't see that much of her."

One holiday break, Judy went to Michael's home to stay with his family. They'd sat down to talk when Michael got on one knee and asked her to marry him. She said yes. So right after graduation—when Judy had finished near the top of her class—she married Michael at Beth El Synagogue, the same place she'd been confirmed during high school.

After a honeymoon to Jamaica, the newlyweds moved to an apartment in New Jersey, spending much of their savings on an upright Steinway piano so Judy could continue performing Chopin. They both found jobs at RCA, with Judy working as a designer of integrated circuits. Even though Judy was one of the company's lone female engineers, RCA recognized her formidable mind. They even offered her a salary higher than Michael's.

After a couple of years at the company, Michael decided to take a hard left turn and go to law school at Georgetown. Judy requested a transfer to RCA's offices in Springfield, Virginia, to stay with him, but she began to wonder if her future really lay with RCA. She didn't feel like RCA knew exactly what to do with her.

Finally, Judy left to pursue her PhD in electrical engineering at the University of Maryland, thanks to a research fellowship from the National Institutes of Health. While she studied, she took on the role of a biomedical engineer at the NIH, using her logical mind to help doctors learn more about the structure of the human eye. It made her optometrist father proud.

It was an intense time for the couple. No more weekends filled with racing, but rather years of study and focus. Judy would spend hours upon hours in the laboratory, peering through a microscope at a frog's exposed retina. But she still questioned whether this was what she wanted to do for the rest of her life. Frustrated, she asked a coworker, "Where do you get your ideas from?" It was as if she was seeking the source of scientific inspiration, which had so far eluded her. His reply: "If you have to think about where ideas come from, maybe you shouldn't go into this kind of science."

Though Judy excelled at all her tasks, she yearned for something she could throw her whole self into.

Eventually, Michael finished law school, and with the future now visible to him, he approached Judy with that possible something— starting a family. But she'd known for years she didn't want children. Her painful battles with her mother had likely brought her to that conclusion.

Her hard line on this topic soon drove a wedge between them. Michael loved Judy but he understood that her life was going in another direction. One day they found themselves in a Howard Johnson's restaurant, divvying up their property. It was as amicable a divorce as one could hope for, and the two would remain friends, making sure to call each other whenever they achieved certain milestones or faced major life crises.

Judy stayed in D.C., throwing herself into her studies and work. During her downtime, she'd visit the beach with friends. She loved getting a tan and soaking up the sun. One day, she mailed a postcard to Canada. It was addressed to Len Nahmi, who was working as a pilot for Air Canada in Toronto. "I'm single," the postcard read. He hopped on a plane to D.C., and Judy was there to meet him at the airport. The reunion sparked the beginning of back-and-forth visits for the two, from D.C. to Toronto. They never became an item officially, but they were back in each other's lives.

And then one day in 1977, Judy turned on the radio and heard an announcement about NASA's astronaut selection. Or she saw an advertisement for it on a bulletin board at work. Or it's possible she heard about the selection from Len, who heard a mention of it on the radio while in his apartment in Toronto. Over the years, the story has changed, depending on which newspaper you read or person you talk to.

But the key detail is that Judy heard about it. One day on the beach, a friend asked Judy what she was doing while scribbling on a piece of paper. "Applying to be an astronaut," she said.

CHAPTER 3

Still Warming Up the Bench

Approximately fifteen years before NASA issued its call for women to join its astronaut corps, Geraldyn "Jerrie" Cobb and Jane "Janey" Briggs Hart sat at a thick wooden table in Washington, D.C., staring back at a mostly all-male panel of lawmakers. Curious audience members filled the cramped room's remaining spaces, giving the proceedings an atmosphere of courtroom drama.

The women weren't on trial, though. They were witnesses at a House subcommittee hearing.

With her curly blond hair tucked behind her face, Jerrie tried to suppress her extreme stage fright as she answered question after question from the lawmakers. Some on the panel sensed her anxiety. John Edward Roush, a Democratic representative from Indiana, looked down on Jerrie and asked her about a statement she'd made prior to the hearing—about being "scared to death" to speak at the event.

"How do you reconcile this emotional statement with the fact that an astronaut must be fearless and courageous and emotionally stable?" the representative asked.

"Going up into space couldn't be near as frightening as sitting here," Jerrie replied. The joke cut the tension and the room erupted in laughter.

The date was July 17, 1962. Jerrie and Janey were there that day to convince both the US government and NASA of the merits of sending women into space. The two women were aspiring astronauts themselves. Accomplished pilots already, they'd undergone a series of tests in the New Mexico desert to determine if they were physically and mentally qualified to handle the rigors of space travel. They proved they were, in fact, capable, fulfilling the same criteria NASA had used to select its first astronauts. And in some tests, they'd even outperformed the men NASA had picked.

But the problem was simple: NASA would not—and could not—accept women into its astronaut program.

———————

WOMEN'S EXCLUSION FROM the astronaut corps had begun almost from NASA's inception, starting with President Eisenhower. Ironically, Eisenhower was never much of a space fan to begin with. He'd scoffed at Sputnik, referring to it as a "small ball in the air," and he never really liked the idea of getting into some sort of "race" with the Soviets, fearing it would add bloat to the federal budget. But even Eisenhower couldn't ignore the Soviet rockets that kept getting launched into the sky. Finally, his administration sent legislation to Congress to turn the existing National Advisory Committee for Aeronautics, or NACA, into what would become NASA.

One key part of this birthing: the new space agency would be a civilian one, rather than part of the military, since Eisenhower actively wanted to keep the exploration of space peaceful. But when it came to selecting those who'd be sent aloft, Eisenhower decided that only military-trained test pilots could handle the job. That view may seem odd, but to him—and to many at the time—such candidates made the most sense. Test pilots were typically healthy and in good shape. They had experience sitting inside the cramped spaces and going breathtakingly fast. And they knew how to take orders, even if those orders sent them to an untimely and fiery death. (Not factored into the analysis was that many male test pilots came equipped with other traits, such as a propensity for booze, fast cars, and extramarital affairs.)

But Eisenhower's decision effectively barred women from going to space in NASA's most formative years. The exclusion *wasn't* attributable to a lack of women pilots. Numerous women had gained piloting experience by the 1950s, and a select few had even flown jets during World War II as members of the Women Airforce Service Pilots (WASP) organization—an initiative that trained women to pilot aircraft within US airspace so that men could be freed up to fly overseas.

But when the war began winding down, fear seeped through the military that the men returning from battle would come home to find their piloting jobs taken by women. As a result, Congress disbanded the WASPs and, for the next three decades, the military banned women from flying its aircraft. It was a perfect catch-22. The only way someone could become a test pilot—and as such steer an aircraft propelled by jet engines—was to join the armed forces. So when the time came in the late 1950s to pick Mercury astronauts, NASA only had men to choose from.

Even if the military *had* been more lenient, other obstacles stood in the way of women who harbored astronaut dreams. NASA's top officials displayed the biases of the era and had been drafting their own preliminary requirements for candidates. They needn't be test pilots, but their candidacy would be regarded favorably if they held degrees in science, math, or engineering; possessed years of experience doing technical work or research; or operated some kind of aircraft or submarine and were willing to accept the hazards that came with spaceflight. They also had to be between the ages of twenty-five and forty and under five eleven to fit inside the teeny tin cans that would be launched skyward. And NASA specified that they had to be "males."

It was a requirement that lined up with the argument mounted by some experts that women who sought employment outside of housework threatened traditional gender roles, and even national security. The newspapers of the time listed jobs in two separate sections: one for men and one for women. One guess as to where the piloting jobs were located.

Yet even if NASA hadn't specified men on that first-draft document, and even if Eisenhower hadn't directed NASA to pick test pilots, women and people of color would have likely still struggled to meet the agency's requirements. Flight experience was great, but NASA simultaneously sought college-educated engineers and scientists, figuring astronauts could learn the ins and outs of the spacecraft they were piloting, and, if something went wrong during a flight, quickly MacGyver a solution. They'd be more than just "spam in a can," a step above flying hunks of human meat at the mercy of their metal vessels.

But not many women and people of color had the kinds of degrees and experience NASA desired. From 1950 to 1960, women represented about 1 percent of all people employed as engineers, and they made up between 9 and 11 percent of employed scientists.

For so many reasons, it seemed that the door that barred female astronauts would remain forever shut. But as NASA began accumulating mountains of data on men in space, a few key people wanted to know how women would fare if put in the same place.

BACK IN SEPTEMBER 1959, Jerrie Cobb had strolled down the sands of Miami Beach, walking toward the crystal-blue waters surrounding the Florida peninsula. It was around seven in the morning, and the sun was just coming up over the horizon. Jerrie, a native of Oklahoma, was in town to attend the Air Force Association's national convention, which was in its second day of meetings. That morning, she walked to the water along with her friend and colleague Tom Harris.

Tom had been her ally at what would become Aero Commander, an aircraft manufacturer where Jerrie worked as a pilot. Originally the company had been reluctant to hire her, despite her extensive experience flying Aero Commander planes. She'd been flying since the age of twelve and had obtained her private pilot's license at age sixteen, the youngest age one could do so. Since then, she'd accumulated seven thousand flying hours, piloting to locations all over the world and ferrying cargo to South America for a previous employer, Fleetway Inc. (A few years later, in 1962, she'd be named "Woman of the Year in Aviation.")

Still, it hadn't been enough to obtain employment with Aero Commander at first. They saw her, a woman, as a liability. So she'd conspired with Tom, a vice president and general manager at the company, to devise a plan. He helped arrange for Jerrie to prove herself by beating an aviation speed record with one of the company's planes, flying a triangle path from Las Vegas to San Francisco to San Diego. It had been one of the toughest flights of her life—with every manner of glitch popping up while in the air—but she'd edged out the record previously held by a Soviet pilot, coming in under his time by twenty-six seconds. It was enough to secure her permanent employment at Aero Commander.

As Jerrie and Tom walked down to the water that day in September, two attractive men emerged from the ocean, having just gone out for a morning dip. Still glistening from their early plunge, the bathing suit–clad men approached Tom, who recognized them and proceeded to introduce them to Jerrie.

"Dr. Lovelace, General Flickinger," Tom said, "I'd like you to meet Miss Jerrie Cobb."

Jerrie instantly recognized their names. General Flickinger was *Brigadier* General Donald D. Flickinger, a flight surgeon with the US Air Force who'd done groundbreaking work in the field of aerospace medicine. And Dr. Lovelace was Dr. William Randolph "Randy" Lovelace II, also a well-known aerospace researcher who'd developed an oxygen mask for pilots to breathe more easily at high altitudes, where the atmospheric pressure and oxygen thinned. He'd first made a name for himself during World War II by bailing out of an airplane at a whopping 40,200 feet to test his mask during emergency scenarios. The stunt ultimately knocked him unconscious in the air and gave him frostbite on his fingers.

But what intrigued Jerrie the most was the men's work in the "space" part of "aerospace." The newly formed NASA had named Randy Lovelace chairman of the agency's Special Advisory

Committee on Life Sciences and tasked him with studying how space travel affected the human body. Soon the Lovelace Clinic, which the doctor ran in Albuquerque, New Mexico, became a proving ground for choosing NASA's first astronauts. At the clinic, Lovelace and his fellow doctors had poked, prodded, and nearly pulverized a group of military test pilots roughly seven months earlier, investigating their fitness for space travel. They'd helped to whittle down a group of thirty-two candidates to seven after days of unique—and invasive—testing. The winners had been introduced to the world as the Mercury Seven.

Fascinated by the men in front of her, Jerrie became even more interested when Tom said the two men had just gotten back from a meeting with space scientists in Moscow. He then revealed to the men that Jerrie was a pilot with substantial experience, noting her various speed records. "She's liable to try for a record in space next," he joked.

That line definitely piqued Lovelace's and Flickinger's interest. They asked Jerrie if she could meet with them later to talk in more detail. She agreed and, a few hours later, she found herself at the Fontainebleau Hotel in Miami Beach, where the two men peppered her with questions about women and aviation: *Were there other women pilots like her? How old were they? How fit?*

Then they got to the point. Both Lovelace and Flickinger were curious to see how women would fare when put through the same testing trials as the Mercury Seven. Since only test pilots could be considered for the Mercury program, all of Lovelace's original guinea pigs had been men. But research had started to suggest that women might be the ideal candidates for spaceflight. They were typically smaller and weighed less, which meant they could more easily stuff themselves inside minuscule spacecraft, and they didn't need as much rocket fuel to get them off the ground. There was also speculation that a woman could better withstand the radiation-filled space environment, and research from the 1950s backed up the idea that women performed better in isolation experiments.

Flickinger had already started planning a program within the air force that would put women through the same astronaut training that the men had been through. And Lovelace had come on board to collaborate. They figured it made sense to test women pilots, specifically, since the Mercury astronauts had all been pilots, too. And after speaking with Jerrie, the two wanted to know if she'd be interested in being a test subject.

With tears filling her eyes, she simply replied, "I would."

FIVE MONTHS LATER, Jerrie lay stiff as a board on a wooden table that rotated back and forth from a horizontal position to a sixty-five-degree angle. Without any straps holding her down, she tried to stay on as best she could without falling on her face. Every so often, a doctor would take her blood pressure while electrodes attached to her body searched for any unforeseen circulation problems that might have been brought on by the oscillations. If Jerrie was ever going to go to space, the doctors wanted to know if her heart and veins could handle being thrown about by a tumbling space capsule.

This "tilt-table test" was just one of dozens of kooky experiments Jerrie underwent in February 1960 at Lovelace's clinic. For about a week, she endured eight-hour-plus days filled with all manner of testing, determined to gauge her balance, her eyesight, her stamina, her lung capacity, and more. The tests had been the same ones Lovelace had performed on the Mercury astronauts the previous year.

The road to testing hadn't been a smooth one. Flickinger's original air force program—which he'd tentatively named Project WISE for Women in Space Earliest—didn't make it off the ground. Right after Flickinger and Lovelace met Jerrie in Miami Beach, another set of researchers in the air force invited a prominent woman pilot, Ruth Nichols, to undergo a series of astronaut tests at Wright Field in Ohio. Her testing wasn't part of any official program—more of a short-lived experiment. But word got out about Nichols spending time in a centrifuge and undergoing isolation tests. When that went public, the air force panicked, fearing that people would think the military was *seriously* considering testing women for space. As a result, the military pulled the plug on Flickinger's initiative so as not to give anyone the wrong idea.

Determined to keep the idea afloat, Flickinger asked Lovelace if he could take on the project at his clinic in New Mexico. He agreed and wrote a letter to Jerrie around Christmas telling her to get her affairs in order before coming to Albuquerque. Fast-forward to February 1960, and Jerrie found herself breathing into strange tubes while riding a stationary bike and getting poked by needles in her hands to test her nerves' reflexes.

When the exhausting week was over, Jerrie received the news she'd been hoping for: She'd passed. Jerrie officially became the first woman to meet the same physical criteria required of the Mercury astronauts. And Lovelace wanted people to know this. In August 1960, he shared his findings with the world, presenting the results of Jerrie's tests during the Space and Naval Medicine Congress in

Stockholm, Sweden. "We are already in a position to say that certain qualities of the female space pilot are preferable to those of her male colleague," Lovelace noted at the time. The news caused a sensation, thrusting Jerrie into the media spotlight. Newspapers and magazines ran stories about this curious "astronette," making a point to call out her 36-26-34 figure, the seven pounds she lost during her stay at the Lovelace Clinic, and the fact that she was scared of grasshoppers. One headline proclaimed "NO. 1 SPACE GAL SEEMS A LITTLE ASTRONAUGHTY."

But through all the absurd coverage, some began to wonder: Could women fly to space too?

Lovelace didn't want to stop at just Jerrie. He hoped to continue testing more women, but he needed money. He turned to a close friend, Jacqueline Cochran—an accomplished pilot who'd been friends with Amelia Earhart. She led the WASPs during World War II and became the first woman to break the sound barrier. Plus, she had money. She was married to Floyd Odlum, then one of the richest men in the world. With her investment, Lovelace's Woman in Space Program had been born, and the doctor got to work recruiting more women to come to Albuquerque. Jerrie handed over some names of women pilots who might make great test subjects, and Lovelace sent out invitations to see if they'd participate. Jerrie herself reached out to her friend Jerri Sloan Truhill, asking her if she'd be interested in participating in tests for a "top secret government project." One young pilot, Wally Funk, reached out to the Lovelace Clinic herself, after seeing Jerrie's story in a magazine.

In the spring and summer of 1961, eighteen additional women arrived at the Lovelace Clinic, often in pairs, to undergo testing.

And just as Jerrie had done, they all underwent the same tests that the Mercury astronauts had endured. Each swallowed a three-foot rubber hose down their throats to test the acidity of their stomach fluids. They had ice-cold water squirted in their ears to test their ability to combat vertigo. Enemas were a daily affair, and the women had to get X-rays of their intestines taken while a tube plugged up their rear ends. Going to the bathroom during those tests meant walking down the hall with a helper—while a tube trailed along with them.

Twelve passed the tests in addition to Jerrie, with "no medical reservations," but the female test subjects' accomplishments didn't stop there. Three of the women—Jerrie, Wally Funk, and Rhea Hurrle—managed to undergo additional psychological testing at the Oklahoma City Veterans Hospital sometime after the Lovelace tests. There they floated in a dark, soundproof tank filled with water, to dampen their senses and create a true feeling of isolation. At the time, the consensus was that hallucinations would start to set in after six hours inside the tank. The women easily lasted for more than nine hours. Doctors eventually had to pull a completely serene Wally out of the water after ten hours and thirty-five minutes. The Mercury astronauts had only been required to last three hours in a dark, soundproof room.

All seemed to be going well for the women in 1961. Jerrie's newfound fame even caught the attention of NASA's administrator, James Webb, who appointed her as a special consultant to the agency. And there were still more tests to perform. Lovelace planned to add more data to his Woman in Space Program by having the women simulate the conditions of spaceflight at the U.S. Naval School of Aviation Medicine in Pensacola, Florida.

Then in August 1961, as most of the women prepared to leave their homes to head to the Southeast, they all received telegrams. The tests had been canceled.

––––––––

JERRIE'S VOICE GAVE away her unease as she read from her prepared statement in the congressional hearing room: "Now we who aspire to be women astronauts ask for the opportunity to bring glory to our nation by an American woman becoming first in all the world to make a space flight."

In making that statement, she was hoping to stoke the representatives' competitive fires. Rumors had intensified that the Soviet Union might launch a woman into space soon, surpassing the US at another critical space milestone. The thought devastated the thirteen women who'd passed Lovelace's tests.

"No nation has yet sent a human female into space," Jerrie continued. "We offer you thirteen woman pilot volunteers." Appealing to the US's competitive instincts felt like the best way to resume the testing.

The Pensacola tests had always been on a shaky foundation. They'd been set up mostly as a favor to Lovelace, who kept them relatively under wraps. But when an official high up in the navy got wind of them, he checked with NASA to see if the agency had approved the tests. Since NASA did not have an official "requirement" for the tests, the navy decided not to lend its resources, and the department pulled the plug.

The cancellation hit the women hard, but Jerrie became determined to save the program. She found help from a politically powerful ally. Of the thirteen women who had passed the tests, one of the subjects, Janey Hart, was married to Democratic senator Philip

Hart. Leveraging her husband's political influence, Janey arranged a meeting between her, Jerrie, and then vice president Lyndon B. Johnson, who oversaw the National Aeronautics and Space Council.

By the time Jerrie and Janey shook hands with LBJ in his office in March 1962, Kennedy had firmly set America on its quest to put a man on the moon by the end of the decade. Anything that seemingly detracted from that goal—like, say, figuring out if women could fly to space too—was just not a priority. Outwardly, LBJ put on a cordial display for the two women, though the truth was that he had no interest in helping. Eventually Jerrie had the right meeting with the right politician, and after enough discussions, a House subcommittee hearing on the "qualifications for astronauts" was set. The event was promised to last three full days, filled with expert testimony.

Jerrie was a pilot, not a performer. Swallowing another three-foot rubber hose may have seemed more comfortable to her than public speaking. But continuing this research meant the world to her, and she dove into the hearing without hesitation.

"The Mercury astronauts had to have 1,500 hours of jet time," Jerrie stated. "We have girls with upwards of 10,000 flying hours."

The women had to walk a fine line that day. They had to make the case that test-piloting experience and engineering degrees, which most of the women didn't have, weren't exactly necessary for becoming an astronaut. But they didn't want to be seen as criticizing NASA or its decision-making processes. She reminded the crowd that America could have it all. NASA needn't wait for the moon landing to start training women astronauts, and such a program wouldn't detract from beating the USSR. The United States already had thirteen women who'd passed preliminary testing. All NASA and the government had to do was let them keep testing.

"I don't think it is too soon to start training women for space-flight," said Jerrie, "and, in fact, we should have started long before now."

The second day, however, everything changed. NASA had arrived at the proceedings. George Low, the director of spacecraft at NASA, had been called as a witness, and he'd brought with him two golden and gleaming astronauts: John Glenn and Scott Carpenter. Almost exactly five months earlier, Glenn had become a national hero when he orbited the Earth in his Mercury space capsule, *Friendship 7*. The men's presence revealed the seriousness with which NASA took this subcommittee hearing—and how determined the agency was to shut this down.

"First, let me preface my remarks by one statement," John Glenn started. "I am not 'anti' any particular group. I'm just pro-space."

But as the questioning continued, the three NASA men made it clear that test-piloting experience was crucial to the astronaut way. There were just certain traits that people acquired through test-pilot training, they argued, and those were necessary for coping with space travel. The men also disagreed with the idea that regular flight hours somehow compared to time spent in a jet aircraft. The level of danger just wasn't the same, in their minds. If anything, they believed that the criteria for astronaut selection should be *more* stringent, not less. And just because the women passed the tests wasn't enough to say they were qualified.

"My mother could probably pass the physical exam that they give preseason for the Redskins, but I doubt if she could play too many games for them," Glenn said.

And as for why women were denied the opportunity to fly jet aircraft—well, that was just a way of life.

"I think this gets back to the way our social order is organized really," Glenn stated. "It is just a fact. The men go off and fight the wars and fly the airplanes and come back and help design and build and test them. The fact that women are not in this field is a fact of our social order. It may be undesirable."

But above all, NASA just didn't see the inclusion of women in the astronaut corps as necessary at that time. They already had a great crop of men to choose from, and that was more than enough. Maybe later, the agency could look at women, the NASA men argued. Just not now, with the moon in their sights. "We don't feel at this time this would be an essential asset for our space program," Glenn concluded.

For the women it was a devastating day of testimony. But despite these blows, many of the lawmakers seemed to side with Jerrie and Janey. A few made it clear that they believed NASA should strive to include women and study how to test women for future spaceflight. Representative Victor Anfuso gave a directive to NASA and its administrator to figure out a program that would include women, so that NASA would continue to receive the congressional support it desperately needed.

Then, after the lawmakers congratulated themselves on a successful event, they ended the hearing a day early. Jerrie and Janey were stunned. No further testimony would be heard from scientists. The women wouldn't get a chance to respond to NASA or provide closing statements out loud. That was it.

ON JUNE 16, 1963, less than a year after the subcommittee hearing, the fears of Jerrie and the other women came true. A twenty-six-year-old female parachutist, with close-cropped curly blond hair, climbed into a hardened sphere that was perched atop a Soviet Vostok-K rocket and blasted off into the sky. Her name was Valentina Tereshkova, and she'd been picked from a pool of five women finalists to become the first female cosmonaut to travel to space. Nearly two years of intensive training culminated in her mission, during which she lapped the Earth forty-eight times. When she returned home, she'd spent more time in space individually than all the US astronauts had combined.

The US may have been locked in a battle of one-upmanship with the USSR, but when it came to sending women into space, this was a race that NASA didn't have much interest in winning. In America, Valentina's flight was mostly written off as a publicity stunt. And perhaps, in a way, it was nothing more than that. The Soviet Union's push to send a woman into space began when officials in the country's space program started hearing about the American women who'd trained at Lovelace's clinic. "We cannot allow that the first woman in space will be American," Nikolai Kamanin, the head of cosmonaut training, wrote in a letter. "This would be an insult to the patriotic feelings of Soviet women." The motivation, as was typical during the Space Race, was political. And, as it turned out, Valentina would be the only female cosmonaut to fly to space for the next two decades.

But the US found other ways to write off Valentina's accomplishment. The American press criticized her lack of qualifications—she was a parachutist, not a pilot, after all. Articles highlighted her plump figure and her infrequent application of lipstick. Soon, rumors began to fly among Soviet and US officials that Valentina had suffered some kind of emotional breakdown during the flight, a vindication to men who rejected the idea of sending women into space. "Their first woman was an absolute basket case when she was in orbit and they were damn lucky to get her back," Chris Kraft, a famous flight director during NASA's quest to reach the moon who became the head of the Manned Spacecraft Center in 1972, said in a radio interview thirty years after the flight. "She was nothing but hysterical while she flew. So how do you know we wouldn't have gotten that situation as well?" (Valentina has long denied she suffered any kind of breakdown.) Chris also argued that NASA didn't want to be in the position of losing a woman during a flight, a PR nightmare.

Throughout the early 1960s, NASA's attitudes toward flying women into space were just too prejudiced to be overcome, and the men of the agency made their thoughts well known. Wernher von Braun, the former Nazi scientist who became instrumental in the creation of the Saturn V, quipped during a speech that "the male astronauts are all for" sending a woman into space, and his friend Bob Gilruth, director of NASA's Manned Spacecraft Center, teased, "We're reserving 110 pounds of payload for recreational equipment." After Valentina's flight, one anonymous NASA official said that the idea of sending women into space "makes me sick at my stomach." Robert Voas, NASA's resident astronaut trainer, gave a speech in 1963 joking that women *would* save weight on spacecraft, only if they left their purses behind. "I think we all look forward to the time when women will be a part of our space flight team for when this time arrives, it will mean that man will really have found a home in space—for the woman is the personification of the home," Voas said during his speech.

The truth is, NASA's reactions simply echoed the sentiments

of the time. The only way people could conceive of women going into space was if they provided some kind of release or assistance for their male crewmates. "The question of man's sexual needs on flights lasting two or three years has to be considered," one expert argued in an article published by the Associated Press. Even Randy Lovelace, who'd been a champion of testing women in space, was motivated by an antiquated view of the future. He envisioned a time when space stations would dominate Earth orbit, and he figured that women would be needed as secretaries and assistants. That was his ultimate impetus for studying women to be astronauts.

At the height of the Space Race, the United States' gendered society just didn't take the idea of women astronauts seriously, and so NASA didn't either. As a result, the feet of US women remained firmly planted on Earth as the men soared above. The women who'd undergone Lovelace's tests slowly faded into the background, never regaining the ability to train for space.

An article in *Life* magazine covering Valentina's launch summed up their fates. "THE U.S. TEAM IS STILL WARMING UP THE BENCH," the headline read.

CHAPTER 4

NASA Catches Up

Sitting in NASA's mission control center in Houston, Texas, Nichelle Nichols punched a couple buttons on the beige console in front of her. A few of the translucent switches illuminated and blinked in response to her touch as she picked up a pencil and scribbled some notes in a notebook. She then turned toward the camera recording her.

"Hi, I'm Nichelle Nichols," she said, smiling into the lens. "But I still feel a little bit like Lieutenant Uhura on the Starship *Enterprise*."

The famous actress's outfit was unmistakable. She wasn't wearing the iconic red dress she'd worn throughout her tenure on the hit TV show *Star Trek*. Instead, she wore a deep blue jumpsuit, the same one issued to all of NASA's astronauts. Nichelle was at the agency's renamed Johnson Space Center in Houston that day in the spring of 1977 to film a promotional video. Her job: advertise NASA's newest human spaceflight endeavor.

Nichelle went on to describe the mythical-sounding vehicle that NASA was in the midst of developing. It was called the Space Shuttle and Nichelle clarified that this white-and-black spaceplane was going to be different from its predecessor spacecraft. Once complete, the Space Shuttle would be conducting regular trips to and from space, making it more like a commercial airliner. One day, it could even be used to build a space station in orbit around the Earth, she said. And because the Shuttle would differ so wildly from NASA's famed rockets, the space agency needed a new crop of astronauts to travel inside it.

"And this would require the services of people with a variety of skills and qualifications," said Nichelle.

NASA was recruiting a new class of astronauts—for the first time since bringing a crop of men into the agency from the air force in 1969. But for this latest selection, the agency's criteria had evolved. NASA still needed experienced pilots who could fly the Space Shuttle into orbit and land the monster gently on a runway back on Earth. But this time around, the agency also wanted scientists, academics, doctors, and engineers, regardless of their time piloting a jet.

And, most of all, NASA wanted women and people of color to apply.

It had been more than a decade since Jerrie Cobb had testified in front of the House subcommittee; more than a decade since the Soviet Union had launched Valentina Tereshkova into space. And since then, no additional women had made it above the stratosphere. America continued to send white men into orbit and then to the moon. And when the Apollo program ended, NASA sent more white men to a space station called Skylab in 1973 and 1974. Another crew of white men launched to space in 1975 on the Apollo-Soyuz Test Project, meeting up with a crew of male Soviet cosmonauts in orbit in a symbolic space détente between the US and the USSR.

But as the same kind of astronauts continued to blast into space, the world below them began to change. In 1963, Betty Friedan had published her book *The Feminine Mystique*, which touted the "problem that has no name." Over three million readers read about the sexism that permeated society, pressuring women to stay at home instead of striving beyond domesticity. Though by that point the embers of a new feminist movement had already been lit, the book ignited the flame. A new wave of feminism spread throughout the country as women advocated for reproductive rights, the ability to take out mortgages and apply for credit cards, the passing of laws against marital rape, and more. Parallel efforts to promote racial justice culminated in the passage of the Civil Rights Act of 1964, outlawing discrimination based on race, color, religion, sex, or national origin.

As waves of change rolled through the sixties and seventies, NASA was slow to keep up. The agency dodged question after question about when women and minority astronauts would fly.

NASA couldn't hide from its failures forever. In 1972, Congress passed an amendment to the Civil Rights Act that strengthened penalties against discriminating on the basis of sex, race, and religion. That same year, NASA administrator James Fletcher began to address the problem openly, hinting that NASA was working on plans for people of color to fly aboard the in-development Space Shuttle.

Around that same time, a NASA employee by the name of Ruth Bates Harris had been trying to instill change at the agency. A black woman, she'd been initially hired to run NASA's Equal Employment Opportunity Office, but when she came on board she was demoted from director to deputy director and given a diminished role. Still,

Ruth repeatedly pressured NASA management to open up jobs to more women and people of color. Frustrated by her bosses' lack of urgency, she took matters into her own hands.

In the fall of 1973, Ruth and two colleagues compiled a report reviewing the state of diversity and inclusion at NASA. It had been an unprompted exercise, but the three felt it was necessary to shed light on the situation. The report was damning, exposing NASA's efforts to recruit women and people of color as a "near-total failure." Minorities made up little more than 5 percent of the entire NASA workforce, according to the report—a representation that was "the lowest of all the agencies in the federal government." Female representation was slightly better at 18 percent, but roughly 88 percent of female employees were in the lowest-paying jobs.

And when it came to female representation in space, the report painted a grim picture: "There have been three females sent into space by NASA. Two are Arabella and Anita—both spiders. The other is Miss Baker—a monkey."

The writers cautioned that if NASA remained on its current path, minorities would only make up 9 percent of the agency by 2001. And they summed up the cost of doing nothing. "An entire generation of people have been cheated from witnessing such [an] experience," of seeing a woman or a person of color go to space.

Ruth and her two coworkers submitted their report in September. A month later, the NASA administrator fired Harris, claiming her to be a "disruptive force" at the agency; he also transferred one of the other report authors. Once the media found out about the firing, the backlash was swift. Numerous headlines decried the decision, a major blunder that came at roughly the same time President Nixon was sacking the prosecutor in charge of investigating the Watergate scandal. Lawmakers questioned the firing, and Senate hearings were scheduled. Eventually, NASA rehired Ruth in a different role, but the damage had been done. The agency had to confront its exclusionary culture.

While NASA licked its political wounds, the agency soldiered on through an intense transition period. With the Apollo program coming to an end, NASA was turning away from deep-space exploration and turning toward a new model of space travel—one in which the Space Shuttle would be the primary ride for all astronauts.

Compared to the Apollo, Gemini, and Mercury spacecraft, the Space Shuttle was something of an anomaly. Built like a plane with a giant hangar for hauling supplies, it contained much more room for people to move around. Crews would consist of up to four, five, and six people at a time. And that meant a wider variety of individuals could be welcomed aboard and tasked with a more diverse set of responsibilities.

During the 1970s, NASA still employed dozens of astronauts from previous selections, but at the time, many were in their late thirties, pushing forty. As NASA eyed flying the Space Shuttle on its first launch in 1979, agency officials realized that they'd need to bring in a new generation to crew the orbiting beast. And eventually officials settled on two types of astronauts to recruit for this new era: pilots and mission specialists.

The "mission specialist" role was a completely new position, with a jack-of-all-trades skill set. Those tapped for the job would oversee the deployments of payloads, manage the Space Shuttle's systems, and conduct experiments in space. And it had the most inclusive requirements that NASA had yet devised for its astronaut corps. Candidates needed to have a bachelor's degree in a science- or engineering-related field (though an advanced degree would be preferred); the ability to pass a physical with more relaxed requirements for hearing and vision; and a height between 5 feet and 6.33 feet.

In March 1976, a group of individuals in NASA's Johnson Space Center sat down at a meeting to discuss these requirements and how the agency would go about recruiting the new astronauts. The group was JSC's astronaut selection board, helmed by NASA's newest director of flight operations, a man by the name of George Abbey. A former air force pilot, George had been an instrumental part of the Apollo program, serving as the technical assistant to Chris Kraft, the director of the Johnson Space Center at the time. George was now in charge of the astronauts, not only picking new ones but assigning them to flights in the years ahead. And he was ready for his crews to look more like America than they had before. He'd selected dozens of individuals at the center to help him in this monumental task, including Joseph Atkinson, the first Black person to serve as a member of an astronaut selection board. The group also included Carolyn Huntoon, a physiologist who'd risen through the ranks at NASA, becoming one of the most senior women at the space center. Chris Kraft had asked her if she wanted to apply to be an astronaut, but she turned him down. Instead, she became the first woman to serve on a NASA astronaut selection panel.

Along with establishing criteria for the applicants, the group discussed how to make sure as many people as possible became aware of the selection. They weighed the merits of advertising at universities, high schools, and Lions Clubs, while sending out notices to newspapers and magazines. These were somewhat revolutionary recruiting tactics for the agency, which typically relied on a "they'll come to us" approach. For one member of the board, it was all too much. Deke Slayton, one of the original Mercury Seven astronauts, was in the room during that first meeting. At one point during the discussion, he stood up. "I want no part of this," he said. And then he walked out.

It was an awkward moment, but not a very surprising one. "There was definitely a feeling that bringing women and minorities into the program was not necessarily a good thing," George said. "So it wasn't universally accepted."

The panel didn't let Deke's outburst deter their efforts. They kept going, formulating a plan for how to get the word out. Then on July 8, 1976, NASA declared to the country: the agency is officially accepting astronaut applications and will take applications postmarked by June 30, 1977.

To boost its recruiting efforts, NASA asked Nichelle Nichols for help. Not only would she film a promotional video but she'd also travel the country, giving talks to students and encouraging a more diverse group of people to apply. "I am going to bring you so many qualified women and minority astronaut applicants for this position that if you don't choose one . . . everybody in the newspapers across the country will know about it," Nichelle recalled saying at the time.

"The Shuttle will be taking scientists and engineers, men and women of all races into space, just like the astronaut crew on the Starship *Enterprise*," Nichelle said during the video she filmed for the agency. "So that is why I'm speaking to the whole family of humankind—minorities and women alike. If you qualify and would like to be an astronaut, now is the time."

The recruitment campaign was fierce. Carolyn and many of the other individuals on the panel traveled the country for months at a time, giving multiple talks and impressing upon people of color and women that this time things would be different. But for some of the women Carolyn spoke to, the idea of a woman going to space sounded absurd.

"I can't imagine any woman who'd want to do that," a young woman told Carolyn during one of her recruiting trips.

"Well maybe you don't want to do it, and maybe I don't want to do it," Carolyn replied, "but there are young women in our country who want to fly in space. And they should be given that opportunity."

IN THE LATE 1970s, the application to become an astronaut was your standard-issue federal employment application form. The document asked for your name, work experience, education, medical history, and three references listing people who could vouch for you. To obtain one of these forms, one simply had to write to NASA's Astronaut Selection Office, and the necessary paperwork would be mailed back.

That's exactly what Rhea did—in a way. When her colleague approached her there in the hospital hallway and reported the news of NASA's astronaut selection, she vowed to learn more. About the only thing she *did* know was that the astronauts trained somewhere in Houston. So she sent a letter addressed to "NASA, Houston, Texas" asking for more information. The US Post Office somehow figured out the right routing, and Rhea received a response telling her how to get the application. Soon after, the form was in her hands.

As Sally read the article in the Stanford paper, a spark immediately ignited inside. She had the right qualifications. All she needed to do was fill out the form. She grabbed a sheet of paper with Stanford's Institute for Plasma Research letterhead and scribbled a quick request for information: "I am a PhD candidate in astrophysics at Stanford University, and am interested in the Space Shuttle Program. Please send me the forms necessary to apply as a 'mission specialist' candidate." She noticed she'd messed up one of her words when writing, so she scratched out the error. But rather than start over, she sent the letter in as soon as possible. A form arrived roughly a week later.

Shannon, for her part, didn't even need to ask for materials after she read about the selection. She already had the standard government form on hand, after applying to so many government jobs over the years. She filled out the form the night she read about the selection in the magazine. Hers was one of the first applications to arrive at NASA.

Anna's was perhaps one of the last. With only weeks until the deadline, she and Bill immediately requested the application and set about getting all the information together. They took some time off, dedicating all their efforts to gathering materials. While poring over the documents, Anna smiled as she wrote down "astronaut" in the box that asked for the kind of position she was applying for. Finally, she sent her application to NASA, postmarked the day before the deadline.

Kathy went through the same routine, writing to NASA for more info, and getting a response that asked if she was really sure this was what she wanted to do. She replied yes, got her packet, and filled it out. She mailed the application in early 1977, promptly forgetting about it as soon as it was out of her hands. She had another expedition to attend to, out in the Atlantic, to complete her PhD work. She dove into that and the thought of working for NASA slipped from her mind.

For Judy, the application didn't end when she'd placed the papers in the mailbox. At that point her work was only just beginning. She'd made up her mind that this was what she wanted, and she decided to throw her whole self into making it happen. She and Len embarked on a small campaign to make her as desirable as possible to the astronaut selection board. She cut her hair to make herself seem more professional. Then, equipped with some of Len's flight instruction manuals, she took flying lessons and picked up a pilot's certificate.

She studied Apollo astronaut Michael Collins's book *Carrying the Fire*, absorbing his experiences and using them as tips on becoming an astronaut. But that didn't feel like nearly enough. She began visiting the National Air and Space Museum in Washington, D.C., to get a better understanding of space history. And once, when Len was in town for a visit, he drew a map of the museum, showing Judy where the offices were. She went to the museum again and walked into the director's office. At the time, the position was held by Michael Collins himself.

"Hi, Mike, how are you?" she said. "My name's Judy Resnik, and I want to be an astronaut."

Rather than eject this stranger from his office, he gave Judy some advice.

"Learn everything you can about the Shuttle program," he said.

———

A TOTAL OF 8,079 people applied to be astronauts for NASA between 1976 and 1977. And the selection panel had the pleasure of reviewing every single application they received. It wasn't a speedy process. JSC's data management system consisted of an IBM Selectric typewriter that would type the same information three times. All the other critical data had to be key-punched directly onto tape that was then run through a processing machine. If someone found a mistake in the data, the applicant's record had to be redone from the beginning. "It was a pain," said Duane Ross, a personnel representative for JSC who helped with the selection.

After tallying all the applications, NASA found that 1,544 women had applied to be astronauts. The panel then got to work eliminating a few thousand applicants who simply didn't meet the criteria. That narrowed down the pool to 5,680 candidates, still a hefty number. But the panel was focused on finding up to forty people. They still had a long road ahead.

With such a daunting number of applicants to weed through, the panel thought it might be a good idea to come up with a points system. Each applicant got assigned a number of points based on how much experience they had and how well it matched the criteria NASA had laid out. Pilots, for instance, would get a certain number of points based on the amount of flight time they had, time spent in a jet aircraft, and combat tours served. Mission specialists would get points based on the types of degrees they earned, work experience relating to their field, and any flight time.

It was a good idea—in theory. "You wind up with everyone with the same score," said Duane. "It didn't work."

Still, the panel soldiered on. They called all the references listed by the strongest candidates, requesting short interviews with friends, family, and coworkers. They grouped people into their disciplines, creating a group for pilots, space sciences, life sciences,

Earth sciences, general engineering, and more. Taking everything into account—from the applications and the points to the reference checks—NASA somehow managed to narrow down the pool even further, eventually settling on 208 candidates, including twenty-one women. Those would be the select few who'd travel to Houston in groups of twenty to go through the final testing and interview process. Now they just needed to call everyone.

IT WAS AUGUST 1977—just another day for Shannon at the Oklahoma Medical Research Foundation—when she received a phone call from someone at NASA. He wanted to know if she'd be interested in coming to Houston for an interview.

"What will the week entail, sir? Who else will be there for interviews?" Rhea asked the man on the other end of the line. She was at work at the Memphis Veterans Administration Hospital, where she'd just started her final year of residency a few days prior.

The man on the phone was Jay Honeycutt, a member of the selection panel and George's assistant. He explained there'd be briefings on the Space Shuttle. There'd also be a physical, so Rhea should bring some running shoes and workout clothes. There'd be a psychological evaluation, too, and an hour-and-a-half-long interview.

"Will there be any other women?" Rhea asked, her mind racing.

"Yes," he said. "Eight of you."

In California, Anna and Bill were at home, going over plans for what was going to be a busy few weeks. Anna's birthday was coming up on August 24, and the week after that, they'd planned to fly out to Windermere, Florida, where they were going to get married at an old house that Bill's family owned. It wasn't going to be a large affair. Just a small gathering with family.

The couple had taken a few days off from their busy jobs working for an ER group in the Los Angeles area. At times, the job could be relentless; both Bill and Anna regularly worked twenty-four-hour shifts a couple times a week. The exhausting schedule had started to take its toll, but the financial compensation had been good, so the two had been able to pay off their loans and save some money to start their future together.

While Anna and Bill went over flight plans for their upcoming nuptials, the telephone rang. Anna picked up the phone. A man from NASA was on the line. She listened to him, her eyes wide. She put down the phone, covered it with her hand, and turned to Bill.

"It's NASA," said Anna, hushed. "They want me to come in for an interview." But there was a catch. She needed to fly in next Sunday to interview for a week—the same week she and Bill had planned to get married. Also, NASA was just calling for Anna, not Bill.

"Say yes, and we'll figure it out," said Bill.

After Anna finished the call they got to work. The Florida wedding was officially off, but rather than push the ceremony to a later date, they decided to move it up. The couple visited the Wayfarers Chapel in Rancho Palos Verdes, a church designed by Frank Lloyd Wright, boasting geometric architecture shaped out of glass and stone. They asked if the chapel had any last-minute openings available and were told that Tuesday at two was free. The couple booked it.

That weekend, they bought a dress, found a photographer, and called their friends in the area to see if they could come. It was too short notice for the families to attend. After a whirlwind four days of planning, Anna became Mrs. Anna Fisher that Tuesday, August 23, 1977. The newlyweds took a trip to San Francisco the next day for their "honeymoon," coming back to San Pedro a day later. Anna then worked two full shifts in the ER, and finally on Sunday morning, she was on a plane to Houston.

She'd packed into her short-term itinerary as much as possible, but there was one thing she just had to skip. Her ten-year high school reunion had been scheduled for the Saturday before she left, and Anna ultimately opted to miss it. Throughout the party, rumor began to spread that Anna Tingle wasn't there because she was interviewing to go to space.

"I'm going to Houston for an interview," Sally told her childhood friend Sue Okie over the phone in September 1977. The news came as a total shock to Sue. Sally had never mentioned wanting to be an astronaut before. "I saw this thing in the Stanford newspaper, and a light went off," Sally told her. In that moment, she realized it was what she wanted to do with her life.

By the time Judy had received her call, her home address had moved across the country. Now toting a PhD from the University of Maryland, she'd accepted a job at the Xerox Corporation and moved to Redondo Beach. There she'd continued her second job of prepping to become an astronaut, by jogging along the beach across the highway from her apartment every morning to get into top shape. Her hard work seemed to be paying off when she received the Houston invite. And her father's words to her were starting to come true. When she'd first applied, she'd called him.

"Daddy, guess what? I've applied to NASA."

"Good, Judy. So you'll become an astronaut," he'd told her.

"Oh, Daddy. There will be thousands of others applying."

"So what? Of course you're going to be accepted."

When Kathy's phone rang in October 1977, it wasn't NASA on the line. Instead, it was a professor at Columbia University. As she entered the final year of her PhD program, Kathy had been applying to postdoc opportunities left and right. And the one at Columbia was particularly seductive. It would entail studying the marine geology of the deep sea, allowing her to fulfill one of her dreams: diving to the ocean's depths in an *Alvin* submersible.

"Hey, are you going to take my postdoc?" the professor asked. He told Kathy she was at the top of his list, but he needed an answer.

"Oh, yes," came Kathy's reply. "Well probably." Did she have something else? he wondered. Not exactly. Kathy explained her predicament: She'd applied to NASA's astronaut program and was waiting to hear back. Coincidentally, the professor had applied to NASA's astronaut program once himself, back in the 1960s, and had even become a finalist before getting cut. Knowing how the process went, he told Kathy to let him know if she knew where she stood. "The probabilities are vanishingly small," he said. "In all likelihood you're going to come take my postdoc."

Kathy figured as much but decided to find out for certain. She searched through the documents she had leftover from the application and found a phone number for someone in Houston.

"Haven't we told you 'no' yet?" the man from NASA said when she called.

"You haven't told me *anything* yet," Kathy said.

He told her to hold on as he checked his notes. After a few minutes, he returned to the phone. "You're scheduled for an interview next month," he said, noting that Kathy should receive a telegram in the mail.

Stunned, Kathy called back the professor, letting him know that she actually couldn't accept his postdoc just yet. She then called her mother.

"What exactly does this mean?" her mom asked.

"It means that when I finish my degree I'm either going two hundred miles up or six thousand feet down."

———————

THE FIRST GROUP of twenty astronaut candidates arrived in Houston on August 2, 1977, marking the beginning of an exhaustive interview process that would stretch through the rest of the year. Every other week, a new group of twenty would show up at Johnson Space Center, fresh-faced and hopeful, ready to tackle the week's worth of activities that NASA's officials had cooked up. While the first two groups consisted of all male pilots, the third, which arrived on August 28, was made up entirely of mission specialist candidates—a cull that included the first eight US women to ever interview as astronauts. Rhea, Shannon, and Anna were among them. Sally, Judy, and Kathy would each arrive on different dates in the next three months.

Each interview week began on a Sunday with an evening orientation meeting, giving the candidates their first opportunity to size up the competition. George and members of the selection board would greet everyone at their hotel—the Sheraton Kings Inn near the center—and give them their schedules for the week ahead. The women looked around at their fellow applicants, a few feeling as out of place as ever. Many of the military men knew each other already; some of the academics did too. But most of the women felt completely foreign, taking little comfort in the fact that many of their peers shared the same credentials: MDs and PhDs. Kathy would later say, "I felt very much out of place and thought, *It beats me; I've got no idea what any of this is.*"

On that first night, George asked everyone for a very simple homework assignment—their first task of the week. He wanted each of them to write an essay on why they wanted to be astronauts.

Late into the evening, the women wrestled with how to verbalize their dreams of launching off the planet. They had no idea what the selection panel was looking for. So, with the absence of guidance, they simply tried to be honest.

"I've been fascinated with space ever since those early telecasts, but never really thought it would be possible for me to become an astronaut—my chances of getting certification as a test pilot appeared pretty slim then," Sally wrote on her sheet of paper.

"I also think it is time that women be allowed into the program—not only because they have a great deal to offer—but because it is time that we knew how they will fare in space—and what special problems they will face there," Rhea wrote.

"I realize that there will be certain significant sacrifices which I must make in both my personal and professional lives in order to become a mission specialist astronaut but those are sacrifices which I thoughtfully and willingly will make if given the opportunity to fulfill a lifetime dream," Anna wrote on Kings Inn letterhead before heading off to bed.

The next day, the trials began—though for Anna, Rhea, and Shannon their challenges started as soon as they stepped off the bus that took them to Johnson Space Center. A mob of journalists accosted them and their fellow female aspirants, shoving cameras and lights in their faces, asking for their thoughts on the program and why they wanted to be astronauts. Everyone wanted to know more about the first women NASA was interviewing to take on this larger-than-life role. The men, who were completely ignored by the media, looked on with expressions ranging from bemusement to jealousy as the women received all the attention. The women offered some quick responses and moved on to their scheduled tasks. They didn't know at the time, but they were getting a small taste of what was to come.

That same day, during part of her physical, Rhea politely answered more of the reporters' questions. The press seemed fixated on two issues: whether there'd be romance in space between men and women astronauts and whether women would be more emotional than men on the voyage into orbit. When Rhea read the papers the next day, she was mortified by the journalists' framing and how they used her words, realizing she had to be more careful about what she said to the press in the future.

Just as the Mercury Seven had to endure a battery of physical tests, so, too, did the incoming Space Shuttle astronauts. The candidates ran on treadmills with breathing tubes over their mouths while physicians monitored their blood pressure; they sat on a rotating chair designed to test their balance and sensitivity to motion sickness. Doctors examined their ears, eyes, noses, teeth, and throats. Finalists sacrificed vials of blood to test for unseen diseases and complications, and no one could escape the dreaded enemas. Cumulatively, there were twenty-four different medical procedures, along with a comprehensive checkup from a NASA flight surgeon.

NASA certainly prioritized physical health, but the agency was just as interested in the aspiring astronauts' psychological fitness. NASA wanted people who could react well under pressure and easily get along with others. So, each of the women had to submit to two different psychiatric evaluations. There was a "good cop" named Terry McGuire, who asked the candidates about their families, their hopes, their desires. He wanted to know if they loved their mothers and if they thought they were organized, while speaking to the candidates in warm and encouraging tones.

"If you could come back as any animal in the world, what would it be?" Terry asked Rhea.

"I'd like to be a dolphin."

"Why?" Terry asked, smiling.

"Well, they're very bright, happy, and playful," Rhea responded.

For Terry, it wasn't *what* the candidates said but the *way* they replied and the words they used. Answering a question quickly or using words like "should" when stating an opinion denoted certain personality traits that helped Terry get an understanding of the person and how they might react to stressful situations.

Terry's foil was Dr. Eddie Harris, the "bad cop" psychiatrist. His questions were uncomfortable by design. He wanted to know if they'd ever contemplated suicide and if they'd experienced abuse as children. Then came his quiz.

"Count backwards from one hundred by seven," he asked Judy.

"One hundred . . . ninety-three . . . eighty-six . . ." the former math major began.

"Name all the presidents from the current president to the year 1900 in backward order."

"Jimmy Carter . . . Gerald Ford . . ."

When someone inevitably got a name or number wrong, Eddie

would declare it loudly and stare at them, waiting for the candidate's response. Again, the test measured their reaction, not their knowledge of US presidents.

At one point during the week, officials asked the women to step inside a "personal rescue sphere," a sack-like ball that stretched about three feet wide. NASA claimed it was a vehicle that future astronauts would use in emergency scenarios, to transport astronauts from a failing Space Shuttle to another Space Shuttle that had come to the rescue. They'd climb inside with an oxygen mask, zip up the ball, and ride inside it through the vacuum of space. The sphere's appearance, however, didn't exactly scream "safety," and the candidates quickly guessed what this test was really about: Can you remain in an enclosed space without freaking out? The women happily curled up in the fetal position inside the ball as they were zipped in. The test lasted only fifteen minutes or so; some of the women dozed off in the darkness.

Everything felt like a test. NASA officials encouraged the candidates to explore the center and talk to current JSC employees—to get a better sense of what working in Houston would be like. But some of the women wondered if this was all a trick. Did NASA *really* want them to walk around? And were agency officials watching and listening in on their conversations with other employees? They approached each task with caution, imagining Big Brother might be nearby. "The implication was if you're strolling aimlessly across the campus versus marching purposefully across the campus, someone will note that and write it down," Kathy said.

The women ultimately did tour JSC, with a few getting more comfortable than others. One day during some downtime, Sally discovered the racquetball court in the gym, where two male candidates were swatting balls against the wall. She asked if the men were interested in a game of cutthroat. They agreed, ignorant of Sally's tennis history, and she summarily obliterated them. "She proceeded to destroy both of us, leaving us defeated sweaty lumps in our respective corners before bowing out as calm and composed as when she'd entered, thanking us and saying she couldn't spare any more time because she had to go running!" Lawrence Pinsky, one of the finalists, recalled. Word spread among the other candidates of Sally's victory, garnering the young PhD candidate some street cred among her group.

As Anna walked around the campus, she started to get a feeling that this place might soon become home. Early in the week she'd asked Bill to come meet her in Houston, and one morning before she embarked on her various tests and interviews, he pulled into the parking lot of the Kings Inn to see her off. "You better go look around and see what you think, because—I don't know, I just have this feeling about all of this," Anna told him. Eventually, Bill *would* come back. He'd secured an interview for November, coming to NASA in the same group with Judy. After realizing they both lived so close to each other in California, Judy and Anna became friends and stayed in touch.

Until this point, everything was more or less pass-fail. The candidates either got the green light after their psych and physical evaluations, or they got a big red stamp on their application. The real test came in the form of the personal hour-and-a-half-long interview with the selection board.

For some, walking into the interview felt a bit like walking into an inquisition. The members of the selection panel sat at a long conference table, facing the candidates as they asked them questions.

Some immediately recognized John Young sitting in one of the seats, a legendary astronaut who was the ninth person to walk on the moon. The women also noted Carolyn's presence, a few thankful to have at least some female representation on the other side of the table.

Perhaps the one positive about the interview was that there was no way for a person to prepare. The finalists hadn't been given any guidance about what would be asked, and that was on purpose. Really, the selection panel wanted to get to know them as they were. The panel only had one question for everyone: "Start in high school, tell us what you did there, and bring us up to now," George would say at the start. Sally described her graduate work with free-electron lasers and her tennis background, while Judy talked about her work with the National Institutes of Health and learning more about the human eye. Rhea and Anna both detailed their medical work. Kathy discussed her deep-sea missions and Shannon talked about her chemistry experience. The women who had flying experience—Shannon, Judy, and Rhea—spoke of their time in the air, and many noted their fascination with the early spaceflight missions.

As the candidates spoke, members of the board would interject questions along the way, tailoring them to each person's career history. The conversations mostly centered around the individual's professional life, but eventually the board members got personal. They asked about times when the candidates overcame struggles and what they liked to do in their spare time. Then came the family questions.

"What if, on the plane going back to Memphis this weekend, you meet the man of your dreams and he asks you to leave your medical career, give up the chance to become an astronaut, and go away with him?" a panelist asked Rhea.

The question annoyed her, and she didn't think it was very

appropriate. Rhea was the first woman to be interviewed of all the twenty-one female finalists, and she figured there were still a few kinks to work out in the questioning process for women. So she answered truthfully. That man *wouldn't* be the man of her dreams if he asked her to do that, she told them.

Anna also tried to be as honest as possible with the panel, including in her choice of outfit: a long green onesie pantsuit with thick-wedged heels, typical of the 1970s. That was her style and she figured NASA knew everything about her already anyway, so she might as well be true to herself. With the same "be who you are" philosophy in mind, when the question of children came up she spoke candidly. "I want to have children, so if that's a factor in your selection, I definitely do want to have children," she told them. She hoped it wasn't a dealbreaker.

As the panel pivoted from one question to the next, Shannon waited for the question about family to come up: "How can you possibly do this job with three kids?" Just two years prior, Shannon had given birth to a son, Michael, making the Lucids a family of five. And years of battling with obtuse former employers regarding the issue of being a working woman and mother had primed her to answer this question.

Ironically, it was never asked. Instead, the panel pointed out: "This job requires a lot of travel. Do you have any problems traveling for work?"

Considering the job she was applying for, it was a funny question. But Shannon knew what they meant. Being an astronaut meant more than just traveling off-world. It would likely require flights around the country to train for missions and meet with scientists, engineers, and contractors.

"Absolutely not. I travel now." She noted that there was a simple arrangement at home. When she traveled for work, her husband, Michael, watched the kids. When he traveled, she watched the kids. It had worked well for the last decade. No reason that responsibility-sharing couldn't continue.

The board couldn't help but get in a couple of devious questions. Some of the early candidates were asked their thoughts on the Panama Canal Treaty, which had just been signed in September 1977. It relinquished America's control over Panama by the year 2000. The goal of the question was to see if the candidates were at least somewhat up-to-date on current events. But the panel also knew that the candidates talked to one another between interviews, so the next day, the question would be about the Suez Canal. Flustered candidates would go on to recite their rehearsed answers on the Panama Canal instead. "Well, it's ours, we built it, and we ought to keep it," one candidate replied, as the panel chuckled.

Everyone agonized over their responses and the reactions from the panel. One candidate told another that he was never going to get picked because of one of his answers.

"They asked me my thoughts on the Vietnam War—and I told them!" he cried.

Others labored over basic questions, such as if they wanted a Coke. Some tried attempts at humor. When one candidate was asked why he wanted to be an astronaut, he responded: "My father was an astronaut. My grandfather was an astronaut. My great-great-grandfather was an astronaut."

"Have you ever had amnesia?" one interviewer asked Sally.

"I don't know. I can't remember," Sally joked.

What the finalists didn't know is that there were no right or wrong answers. But there *were* certain traits the panel was looking for. Above all, the selection board wanted team players. The Space Shuttle crews would be bigger, with more coordination between the vehicle's inhabitants, which meant people would need to rely on each other to a greater degree. That meant big egos were summarily dismissed.

The panel also sought people who could be flexible, with interests in more than one field. Sure, getting a master's or PhD was impressive, but the board looked for hobbies or side projects that indicated the candidate didn't just focus on work alone. Say, for instance, an interest in flying, playing classical piano, or playing amateur tennis. This was key, since Space Shuttle crews would wear a diverse array of hats during their missions, and they'd need to understand various disciplines—from engineering and rocketry to astronomy and Earth science.

But most of all, the panel tried to gauge if the astronauts really *wanted* the job they were signing up for. The selection board spent a large chunk of time just explaining the Space Shuttle itself and what an astronaut's life would entail. "We wanted to make sure that they understood what the job was because a lot of people had an impression of what astronauts do—and it's quite different than what the astronaut job really is," said George. However, the dangers of the job weren't exactly emphasized during the discussions.

Ultimately, the job would require patience. The Shuttle's inaugural flight was scheduled for 1979, with many of the first crews made up of astronauts already in the space program. It'd be years before this new class of astronauts would get a chance to sit in the Shuttle's seats. That meant most of their work would be on the ground, running experiments, doing simulations, learning software and new technologies, and making public appearances for the agency. Going to space was actually a small part of the job, and the panel wanted people who understood that.

"We didn't want to go through the time and expense that we did to select people for them to get there and say after a few weeks, 'You know, I don't really like this,'" said Carolyn. A few of the finalists wound up telling the selection panel that this life wasn't for them.

To truly uncover a candidate's personality, there was one final tryout: a barbecue dinner at Pe-Te's Cajun BBQ House. For George, this was still a critical part of the selection process, as it gave him the chance to see the candidates in a casual setting. The women weren't sure what to think about this intimidating man, but they could sense he was an important person to impress. Current astronauts came to meet the hopefuls, passing on their thoughts to George and other members of the panel. It was the last bit of work needed to really know who'd best suit the program.

After seven days of intensive work at the Johnson Space Center, each of the women flew home—to California, Nova Scotia, Tennessee, Oklahoma. If nothing else, the six women figured the experience would be something they'd tell their kids about someday. "Hey, did you know that I interviewed to be an astronaut once?" But secretly, they yearned for the day their telephone would ring and a NASA official would be on the other end.

But for now, they had to wait.

CHAPTER 5

Are You Still Interested in Coming to Work for NASA?

"So can you tell me anything?" asked Judy on the other line. Duane Ross laughed. "Nope, they haven't released it yet."

This had been the fourth phone call Duane had received from Judy over the last two months. Since her interview in early November, she'd managed to find Duane's number and called him a few times every other week to see what information she could pull out of him. For Judy, waiting to find out her fate had been an excruciating process, and she tried every method she could to remove the uncertainty.

"Well, I'm going on a backpacking trip, and I'll be out of contact," Judy told Duane, noting she'd be gone for a couple of weeks. She told him she was worried that if NASA called and didn't reach her, management might change their minds and go with someone else.

Duane admired her persistence but was committed to saying nothing. He assured her that she could go on her backpacking trip, though.

The truth was, NASA was getting close. It was late December, and the committee was just waiting on final approval of its choices. At the beginning of the month, the panel had submitted the names of forty candidates—twenty pilots and twenty mission specialists—whom they wanted for the new astronaut class. But NASA's administrator at the time, Robert Frosch, had questioned the need for so many pilots with so many astronauts still waiting to fly. So George and his team went back to their list and did their best to shave off five pilots, all white men, to shorten the list to thirty-five. It took some time to find the right people to remove. Everyone was supremely qualified.

As five hopefuls unknowingly got added to the cut list, the finalists waited in anguish for their calls. Candidates who'd met during their group interviews would call each other every couple of days to see if there was any news. "Have you heard anything?" one would ask frantically. "No, nothing," the other would reply. They'd been told the decisions would be made in early December, but December came and went. Still no word.

The women had all returned to their normal lives in the meantime. For Shannon, that meant continuing her work in the lab, while caring for her family. Judy chugged along at her new job at Xerox. Anna and Rhea toiled away during long shifts in the ER and at the hospital, while Sally and Kathy each worked toward the last year of their PhD programs, their theses looming in the distance. It was life as usual, though NASA was still on their minds.

In late October, Sally took a break from her graduate work to trek down to Edwards Air Force Base in California's Mojave Desert. There, she hoped to get a glimpse of NASA's legendary Space Shuttle in action. That month NASA was conducting its last few landing tests using a Space Shuttle prototype dubbed *Enterprise*. And people actually had to be in the vehicle to complete the test. Astronauts Fred Haise and C. Gordon Fullerton rode inside the test Shuttle's cockpit as the giant vehicle ascended on the back of a Boeing 747, high above the California desert. Then, at just the right altitude, the 747 separated from the Shuttle and flew away, leaving *Enterprise* to plummet back to Earth all on its own.

With a joystick in hand, the two astronauts guided the Shuttle down through the clouds to a runway in the sands of the spaceport, practicing what a landing might be like when the Shuttle came back from space. Sally watched from the ground, along with some other candidates who'd interviewed at NASA, as Fred and Gordon navigated the spaceplane to the intended landing zone. But when the vehicle touched down on the concrete, the spacecraft's massive tires bounced into the air, causing the whole thing to wobble and one of the wings to tip. *Enterprise* bounced a few more times before coming to a final stop.

It didn't look very reassuring, prompting one air force officer in attendance to shout, "What the hell was that?" Prince Charles had even been among the crowd of onlookers. Not a great showing with royalty present. As for Sally, the test landing certainly didn't make her feel particularly secure about becoming an astronaut, either.

"Up until this morning I wasn't afraid at all," she wrote in a letter to Molly, who was living in New York by then. "It never entered my mind that there were actual risks involved. But seeing that landing this morning, knowing that they couldn't pull up for another try—that scared me."

January brought a new year—1978—and still no word from NASA. The finalists watched their silent phones, willing them to jingle. The agony of not knowing was at times a nearly intolerable weight.

After a couple weeks, the phone *did* ring for one candidate in January. While working at the VA hospital, Rhea received a call from, of all people, a reporter. And not just any reporter, but Jules Bergman at ABC. He'd made a name for himself covering the space program and the Apollo flights, though that status was lost on Rhea, who wasn't an avid news watcher. She asked him why he wanted to talk to her. He explained that he'd learned the astronaut selection would be announced soon and he wanted to talk to some of the women who'd interviewed for the job. Thinking the request was bizarre, Rhea agreed, and they set a date to meet at the hospital on Sunday, January 15.

That day, Rhea broke away from her patients to meet the ABC camera crew in front of the building. They set up their shots in one of the exam rooms, and the interview got underway. Everything proceeded normally, with Jules asking the basics about why Rhea wanted to be an astronaut and about her background. Then he dropped a bomb.

"Would you believe you've been picked as one of the first woman astronauts?"

Uh, is this a joke? Rhea wondered. She thought it might be some kind of hypothetical question.

"You're kidding?"

"No, I'm not kidding."

"How do you know that?" she asked.

"Through very good sources."

Rhea fought back her excitement, thinking she'd look ridiculous if his prophecy turned out not to be true. Smiling, she told him she'd still be surprised if she got the call.

"Well, I can guarantee you: you have been selected," Jules said.

"Well, I'm very excited then," Rhea said, laughing.

"They'll be calling you—early in the morning."

"Well, that's what I want to do with my life, and it just . . . it'd be very thrilling to hear that," Rhea said.

The interview ended abruptly, and Rhea sat there, unsure of what had just happened. As the crew got ready to leave, Jules assured Rhea his sources in Houston were sound. The call was coming, he said. The ABC team left a dazed Rhea in the hospital's exam room.

Rhea wasn't the only one whose senses were tingling. Across the country in California, Judy called Anna the same day of Rhea's interview. Anna had gotten to know Judy through Bill, and they'd stayed in touch after everyone's interviews had wrapped up. It had been easy since the two lived so close to each other in the Los Angeles area. Judy felt like something was up, she said over the phone. She'd been getting calls from reporters and wanted to know if Anna was hearing anything, too. Sure enough, Anna was getting the same weird requests. Roy Neal, a space correspondent for NBC, had called Anna that same day for an interview. The two wondered, *Does this mean we've been picked?* "We kind of had an idea that something was up, but you didn't want to get your hopes up too much," Anna said.

The women went to sleep that night, some too antsy to shut their eyes, wondering what the next day might bring.

Bright and early Monday morning, January 16, 1978, George showed up at his office in Building 1 at the Johnson Space Center—a large rectangular structure of alternating windows and concrete that housed the offices of senior staff. When he got to his desk, he reached quickly for his phone. The sun hadn't even come up in Houston. He had thirty-five calls to make that day, to people located all over the country—and some even out of the country. And he didn't have much time to waste. NASA was going to send out a press release that afternoon, announcing the selection of the newest class of astronauts. The news was bound to make a splash. Among the thirty-five people selected, three were African American and one was Asian American, making them NASA's first astronauts of color.

Also within the group: America's first six women astronauts.

The first number George dialed that day was a Canadian one. The ringing telephone jolted Kathy Sullivan and her roommates awake in their apartment in Halifax, Nova Scotia. It was around six thirty in the morning and the sun had barely risen. *Who could be calling at this hour?* Kathy thought, wondering if something awful had happened. Her roommate picked up the phone, listened, and turned to Kathy in awe. "It's somebody from NASA."

She passed the phone to Kathy, who heard George's voice on the other end.

"Are you still interested in coming to work for NASA?" George mumbled, with the same calm inflection as someone asking for a small favor. This question would be George's opening line throughout the day. He truly wanted to know if the people he'd picked still wanted the job. Otherwise, he didn't want them to come down to NASA and waste everyone's time. "There was at least one who told me no," George explained about a finalist he called in a future selection round after 1978. "She'd decided it wasn't really what she wanted to do."

That wasn't the response George received from Kathy that morning. "Yes!" she cried, completely taken aback.

George made his way from Kathy's time zone and moved westward, calling those located on the East Coast first. Soon, he was dialing Tennessee. Rhea was back at work that morning, still buzzing from her interview the day before, wondering if she'd get a phone call. Sure enough, as soon as she pulled up to the hospital, her beeper went off. She checked with the hospital's operator, who informed her that she had a call from NASA.

"Are you still interested in coming to work for NASA?" George Abbey asked when she got on the phone. The women at the hospital's reception desk stared at Rhea eagerly while she took the call. They'd all known she was waiting to hear about her selection. "Yes, sir!" Rhea managed to blurt out, feeling as if she was about to dive off a cliff.

A few minutes later, a coworker at the Oklahoma Medical Research Foundation popped into the room where Shannon was working to tell her that someone was trying to reach her on the phone. Shannon went down to the dimly lit basement where there was a pay phone. After putting in the right change, she dialed the number she'd been given.

George asked his standard question and Shannon sang out, "Of course!," realizing her lifelong dream was coming true. Later that day, she gleefully broke the news to her husband, who was thrilled. The couple then tried to explain her new job to their children, who didn't quite understand what it all meant. "Your mommy might be like Mr. Spock in *Star Trek*," her husband said, the best explanation Michael could come up with.

By the time George made his way to the West Coast later in the morning, many of the candidates were still fast asleep. Sally's phone rang at around 5:00 or 6:00 a.m., way too early for any decent

Californian to be awake yet. Groggily, she picked up the phone to hear George on the other end, asking if she was "still interested" in coming to NASA. Thinking this might all be a dream, she replied "Yes, sir!"

Right after she hung up, she rang Sue Okie. "Hello, this is your friendly local astronaut," Sally said to her friend, beaming. From then on, that would be the way Sally greeted her on the phone.

Judy had just walked out the door to go to Xerox when she heard her phone ringing in her apartment. Wondering if it might be The Call, she sprinted back inside and picked up the receiver.

"Are you still interested in coming to work for NASA?"

She accepted, thrilled that her hard work had paid off (and relieved she'd turned around that morning). When she put the phone down, she called her father and friends, many also just starting their day.

"I did it," Judy exclaimed immediately.

Just a few miles from Judy, the phone rang at Anna and Bill's home. An NBC camera crew was standing with them, their cameras trained on the couple as they stood in their kitchen. They'd asked to film the couple while they received word from NASA. Bill picked up the beige telephone, turned to Anna with a smile, and said, "It's for you."

"Are you still interested in coming to work for NASA?" George asked. Smiling, Anna gleefully replied, "Oh, you know I am! I don't know what to say except thanks so much. We've been thinking about it and thinking about it. I hope I can do a good job for you all."

After Anna accepted, she handed the phone back to Bill, who then became the only person to ever receive a rejection from George himself. All the other candidates who didn't make the cut heard from other members of the board like Jay Honeycutt. If George was speaking to you directly, it was good news—except for Bill that morning. Bill and Anna had prepared for every kind of outcome they could think of. And they assumed that, if only one got picked, it would be Anna, since she possessed more graduate experience in physical sciences. Bill handled the call with grace and embraced Anna when he hung up.

All the women had received their phone calls by the time NASA issued its press release that afternoon. In Washington, D.C., Robert Frosch and other high-ranking agency officials called a press conference at NASA Headquarters to discuss their selections. One reporter asked why it had taken NASA so long to finally choose women to fly to space. "I think that in the last few years in the United States that—because of the Women's Movement frankly—that women are much more qualified," Chris Kraft, the head of JSC, said. He added: "And in this particular case, we found a very large number of women that were as well qualified as the men. That's rewarding to us. And I don't think we're going to have any problems in dealing with the matter once we get going."

Another reporter apologized for asking what he referred to as a "male chauvinist question," but he pointed out that Shannon Lucid had three children. Did her motherhood help or hurt her chances of becoming an astronaut? "We did not take that into consideration at all," Chris replied.

By that afternoon, the dam had burst. Now, the entire country knew about the Six, and soon each of their phones began shrieking nonstop as reporters tried to reach them for interviews. Throngs of media professionals arrived at Rhea's hospital, forcing her employers to set up a press conference. Stanford scheduled an afternoon press conference for Sally as well, where she was surrounded by media shouting questions at her.

"What's the highest you've ever been before?" one yelled.

"Aren't you afraid of being in orbit with all those men?"

Later that day, TV crews arrived at Sally's house, wanting to film her as she ran for her daily workout. Connie Chung, then a local reporter in California, showed up at the Fisher home to interview Anna. At one point, Judy and Anna decided to combine forces, agreeing to do a joint interview for a *Los Angeles Times* reporter so that they'd have each other as support. "I think that when there are as many women astronauts as men it won't be a novelty any longer and the interest will naturally fall off," Judy told the paper. "NASA should have had women involved with the space program many years ago." Reporters desperately wanted to know if Bill was bitter that Anna had been picked and not him. "It's fantastic that Anna was chosen, and I feel no resentment," Bill would later tell a reporter.

It was both an electrifying and overwhelming day for the Six as they tried to satisfy an insatiable press corps without any training whatsoever. "One day I was a doctor working in medical training," Anna said. "The next day I was an astronaut. But nothing had happened in that twenty-four-hour period. I didn't know any more than I knew the day before . . . So that was a baptism by fire." When the dust finally settled, Bill took Judy and Anna out on the town for a nice dinner to celebrate their selection. Afterward, they found themselves back at Judy's Redondo Beach apartment for a night cap. The two women stood on her balcony overlooking the beach, taking in the day they'd just had. Then they both looked up at the sky.

"Can you believe this is really happening?" Anna whispered to Judy in awe.

Over the next few days, the women started to understand just how many eyes were focused on them. Requests rolled in from *The Today Show* and *Good Morning America*, with producers offering to fly the women to New York City for whirlwind visits and interviews. Newspapers printed close-ups of their smiling faces on their front pages. The major networks all asked to visit the women at their homes and at work, to get a peek inside their daily lives.

A CBS News crew profiled Rhea and Shannon in their home-towns, giving viewers a glimpse of Rhea at work during surgery and Shannon pipetting chemicals in the lab. Rhea told the crew that her male friends made jokes that she and the other women would be serving coffee on the Shuttle instead of doing actual work. The crew also wanted to know how Shannon would juggle her responsibilities as a mother and an astronaut.

"Well, I've always worked and I've always put in a lot of hours," Shannon replied. "My oldest daughter, her great discovery when she went to first grade, she came home one day and said, 'Guess what, Mommy? Did you know that some mothers stay home all day and never go to work?'"

As fun as the attention was for Shannon, it soon started to feel a bit invasive. The CBS crew had filmed her at home while she made dinner. After the interview, she decided she didn't want the press coming over to the house or interacting with the children anymore. "I mean, this was *my* job; it wasn't a family job," she said.

But the scrutiny was only about to intensify. NASA sent tele-grams to each of the new astronauts, informing them that they'd need to come to the Johnson Space Center from January 30 to Feb-ruary 1 for an orientation session. The visit would include briefings, but most important, there'd be a press day when the astronauts would be officially revealed.

Of course, everyone complied. All thirty-five of the new recruits boarded planes or hopped in their cars and made their way to Houston.

On the last day of the month, a mixture of NASA employees and members of the press gathered inside JSC's Teague Auditorium, filling up the room's dark floor seats. They faced a raised stage covered in burnt-orange carpet. Two staggered rows of molded chairs stretched across the platform—all empty. NASA's redesigned logo—a sleek, bright red typeface known as "the worm"—adorned the back wall. As camera bulbs flashed and reporters scribbled in their notebooks, Chris Kraft stood at a brown podium, welcoming everyone to the event. Then he began reading out the names of the new astronaut class, officially known as NASA Astronaut Group 8. One by one, each candidate timidly walked out to one of the thirty-five seats as his or her name was called while a photographer onstage snapped a close-up of their face.

The first person to take the stage was Guion "Guy" Bluford, one of the three Black men NASA had selected. The other two, Frederick Gregory and Ronald McNair, would soon follow. Both Guy and Fred had been pilots with the US Air Force, while Ron had graduated from MIT with a PhD in physics. NASA had selected Ron and Guy as mission specialists, and Fred as a pilot. Also among the group was a man named Ellison "El" Onizuka. A US Air Force test pilot and engineer, he hailed from Hawaii and was the first Asian American selected.

"Anna Fisher," Chris called out after a few names had been

read. Anna, wearing a brown skirt suit with a tie, became the first of the Six to take the stage. She took her seat cautiously, her face emotionless to mask her nervousness. One of the men next to her leaned in and made a joke, and Anna broke out in a smile. Eventually, Shannon followed in a white blouse and black trousers, sitting on the end of the top row. Judy and Sally came next, sitting next to each another in the bottom row. The two, both clad in white, whispered to each other, likely feeling lucky to be placed next to another woman. Rhea and Kathy—with last names closer to the end of the alphabet—brought up the rear, flanked by their male colleagues.

With each of the chairs filled, the country got its first look at the most diverse class of astronauts NASA had ever selected. Though the group was still dominated by twenty-five white men, it was the closest NASA had ever come to picking astronauts who reflected the true makeup of America. It was a promising start.

To round out the event, Chris introduced the group as "a great bunch of guys," getting laughs from the crowd. "Well, girls are called guys these days," he chimed in.

After the press snapped a few pictures, NASA officials announced that the astronauts would be available for interviews and herded everyone to the nearby public affairs building. There, the thirty-five new astronauts waited to be called to talk to the press. And perhaps for the first time in American history, the white men found themselves completely ignored. Almost immediately, NASA officials whisked away the women and minority astronauts, while the white men sat around, waiting for a little action. It was just the beginning of what was to come. The lopsided press attention would ultimately lead the astronauts to refer to themselves as "ten interesting people and twenty-five standard white guys."

Just before the women confronted the ravenous press corps, they met with the sole woman who'd served on the selection board. Carolyn Huntoon, then chief of the Biomedical Laboratories Branch at NASA, gathered the Six together and gave them a pep talk. Tall with short dirty blond hair, Carolyn understood what they were about to face. The media would want to know as many details as possible, and how the women answered these questions would spill over to the rest of the group. If one woman divulged all her secrets, the press would try to get that same level of detail from the others. Their answers would also provide lasting impressions about how women astronauts should be perceived. Carolyn prepared them to get all manner of questions about their relationships, their makeup habits, their exercise regimen, and more. Now was the time, she said, for the Six to stand united and decide what they were comfortable talking about—and what was off-limits.

"When someone gets the Nobel Prize, all of a sudden people start asking them questions about 199 other things besides their own field of interest," Carolyn later recalled. "We ran into a similar kind of thing with the astronauts."

In those brief moments ahead of their first interviews, the women strategized. They decided that they'd be open—to a degree. They were taxpayer-funded employees after all, and facing the public was part of the job. But the media didn't need to know every detail about their lives, and they resolved to keep their private lives private.

Then it was off to the wolves. Each of the Six broke apart to do separate interviews with reporters.

Journalists asked whether the women felt different from their male counterparts:

"We all look at ourselves as just one of the guys, one of the astronauts, not as men or women," Judy told a reporter at the *Houston Chronicle.*

They asked who wanted to be the first American woman in space:

"I just want to be a person going into space," said Anna. "I don't really think it is important who the first woman is."

They wanted to know why it took so long for women to enter the space program:

"I think it had to do with the fact that there weren't as many qualified before and also on some of the early flights it had to do with strength," said Rhea.

With the first round of interviews complete, the women emerged from their respective conversations, heads spinning. That's when they found themselves in the hallway again. In that moment, one of the Six recognized an opportunity and shouted: "Potty break!" The Six swiftly gathered in the nearby ladies' restroom. It turned out to be the only haven for them in the building. Most of the reporters and NASA employees handling them were men, so the bathroom offered the day's first bastion of privacy. There, next to the empty toilet stalls, the women all exchanged their experiences.

"Who did you have?"

"What did he want?"

"What did you tell him?"

The women gathered intelligence on each of the media personalities, rehearsing answers to the questions they knew would be asked. Then they scattered, like footballers from a huddle, and ran off to meet their interrogators. The bathroom parleys became a routine that day. Not only did they permit strategy checks but they also provided the first chance to form a united front.

The day was equal parts interrogation and photography session. The women posed together for pictures, famously standing around a small mockup of the Space Shuttle, placing their hands on it as if it were a beloved pet. After the press had exhausted one angle, they asked the women to move around, while someone in the back shouted, "Now let's have a high kick!" When the picture session ended, Kathy quipped, "We'd like a picture of the white males, please."

Perhaps unsurprisingly, those "white males" were long gone by that point. Their jobs basically concluded about five minutes after the presentation. Many had already left to get lunch, while a couple cut out early to make use of the NASA gym. As those men relaxed, the women and minority astronauts found themselves talking to the media late into the night, finally leaving NASA after hours of interviews.

As much as they tried, the women could only control their narratives so much. In the weeks and months after the selection, the press used their own colorful terms to describe NASA's new astronauts. The *Midnight Globe* referred to them as "the Glamornauts" and NASA's "eye-popping space gals." When introducing the women on TV, one anchor read off their names one by one, followed by each woman's marital status and emphasizing the ones who were single. Various articles referred to them as "girls" or "ladies in space," and diligent writers made a point to include ages, heights, and weights in their descriptions. *People* devoted an entire spread to the Six, with the subhead: "YOU'RE GOING A LONG WAY, BABY."

Late at night on America's television sets, Johnny Carson also noted the Six's selection after NASA's announcement, using it as an opportunity for a joke. "I think that's great; that ought to be interesting," he told his audience. "Imagine a woman astronaut two million light-years out in space. She says, 'My God, I forgot to leave a note for the milkman.'" The audience tepidly laughed at first, following it up with some loud groans. Still, he dug in. "Can you see the male astronaut telling the female astronaut: [he] says, 'Honey, we're going to dock tonight.' And she says, 'Not tonight, I have a headache.'"

But a few in the media did convey the selection's historical significance.

"ASTRONAUTS HURDLE SEX, RACE BARRIERS," one headline read. Another: "MEN ASTRONAUTS TAKE BACKSEAT IN NEWEST CREW LINEUP."

The last day in Houston, the thirty-five astronauts gathered for a celebratory dinner, hosted by the man who was about to become their boss: George Abbey. The selectees mingled and introduced themselves to the people who'd someday become their crewmates and allies—and, in some cases, spouses.

A tall redhead named Steve Hawley introduced himself to Sally Ride and the other doctoral students. Originally from Kansas, Steve had been in Chile studying astronomy at the Cerro Tololo Inter-American Observatory when, on a whim, he'd decided to apply. He felt comfortable around the other graduate students and postdocs, hanging out with them the most throughout the night. And he took particular notice of Sally.

But the military folks and the civilians did break down their barriers.

"Hello, I'm Rhea," said the blond Tennessean that night, extending her hand to two of the men she had recognized from the presentation.

"Hi, I'm Mike Coats, and this is Hoot Gibson," one of the men replied, pointing to his sandy-blond friend with a thick mustache. Hoot immediately noticed Rhea's forwardness. Back then, "I'd say maybe twenty percent of women would hold their hand up to shake your hand," Hoot remembered of the 1970s. Rhea shared pleasantries with the handsome men in front of her—both navy pilots. She slightly deflated when she learned they were both married.

Judy also introduced herself to some of the men that night. She met John Fabian, a remarkably tall aeronautics engineer with the US Air Force, as well as Frederick "Rick" Hauck, a navy pilot and engineer who'd graduated from MIT. And then there was Mike Mullane, an engineer and weapons system operator for the air force, who'd been a major space buff throughout his childhood. Though he didn't say it, that night he viewed the women and the other civilians at dinner with suspicion. He wondered what they were even doing there. To him, these baby-faced graduate students didn't have the right life experience to fly to space, not the kind that military astronauts had accrued during combat tours in Vietnam. And his thirty-two years had yet to prepare him for working with women. He wasn't the only one present with that mindset.

Regardless of what people were thinking, the night was one of celebration. Everyone was buzzing after meeting the people they'd be flying with. This new chapter of their lives, one of space exploration, was just beginning.

The group was told to report for duty first thing July 10.

CHAPTER 6

Jet-Setting

I n the summertime, the south of Texas is more akin to the surface of the sun than the freezing-cold vacuum of space. As soon as you walk out the door, the thick humid air streaming in from the nearby Gulf of Mexico wraps itself around you and smothers you in an inescapable, blistering hug. Just taking a few steps outside is enough to trigger rivulets of sweat.

Houston's oppressive heat is often at its worst in July, the month that NASA's new Shuttle astronauts had to report to Johnson Space Center for training. By that point, the Six had all made their cross-country moves, with Judy getting the very first taste of living in the deathly Houston heat. After the presentation in January, she went back to California, quit her job, packed her things, and came out to Houston a couple months before she was officially needed. She'd made the move alone, with Len remaining in Canada as a commercial pilot. The early relocation had given her access to NASA's gym, and she'd even been able to fly on NASA's fleet of jets. She'd also struck up a friendship with two other astronauts—Rick Hauck and Jim Buchli—who'd also come to Houston weeks earlier than required.

Kathy and Sally had gone back to school to defend their theses. Eventually, they traded in the blistering cold of Nova Scotia and the cool Northern California air for Houston. Sally had road-tripped through the Southwest, bringing her physicist boyfriend, Bill, with her for the ride. Thanks to help from Carolyn Huntoon, NASA made the rare move of giving Bill spousal status at the agency, despite Sally and Bill's not being married (still a fairly big taboo in the 1970s). NASA paid to move Bill with Sally to Houston, and he'd eventually get a badge that permitted him access to various NASA properties and events.

Before they embarked, Sally had warned Bill that NASA was going to be very different—a place filled with crew cuts and military types instead of hippie graduate students. So, prior to the big move, they'd gone up to Muir Woods, north of San Francisco, where they bought a rotating bookcase made from California's famous redwood trees. That way, they'd have a piece of Stanford in Texas with them.

The Six had secured home loans and mortgages in the Houston area. And as the press had been so curious to know, both Shannon's and Anna's husbands had managed to find employment there. Michael eventually landed a job at the Shell Oil Company, while Bill Fisher continued his ER work. Bill also picked up a major side gig: he signed up for graduate courses in engineering at the University of Houston and trained for his pilot's license—to improve his chances of getting picked to be an astronaut during the next selection round.

For everyone, the new steamy climate took adjusting to—even for the Lucids, who were used to a hot environment after living in Oklahoma. One day, Shannon came home and found the kids inside, languishing on the couch. She asked why they weren't playing outside. "It is just *so* hot," they cried.

119

Bright and early on July 10, a Monday, the Six all made their way to NASA, the first day of their new jobs. That morning, Anna zoomed to work, hardly believing she was driving to her first day of being an astronaut. There, she'd see Rhea pull into the parking lot. Eventually, Rhea would pop out of the driver's seat of a Corvette Silver Anniversary whenever she'd arrive at work. The car would quickly make an impression on the NASA campus.

The Corvette was a false tell if one was guessing the new astronauts' actual salary. When NASA had sent out its recruiting call, it had noted that salaries would range from $11,000 to $34,000, depending on achievements and experience. The Six had been quoted right in the middle. Sally's initial salary began at $21,833, while Rhea told a reporter she was making $24,700. It was serious money for the graduate students, who hadn't had salaried jobs before, but Anna's take from working in the ER was effectively cut by a third. As a single woman, Rhea's salary was so low that she had to ask her dad to cosign a home loan. The compensation didn't matter to the Six, though.

"I'd pay them for this job," said Rhea. "We all agree on that."

That Monday morning, the astronauts took in their new workplace: NASA's Johnson Space Center. In front of them stood a crop of beige and white rectangular buildings spread out on 1,620 flat acres of once pastoral land. The campus wasn't actually in Houston's center, but about twenty-six miles southeast, near the town of Clear Lake City.

When NASA began construction on the Houston campus, the agency picked the architecture that screamed "government complex" the loudest. Every structure was designed in the quintessential Brutalist style common to institutional buildings. All curves were absent; boxes and right angles abounded, constructed out of mostly concrete and glass. The most eye-catching feature of the property was a full-scale Saturn V rocket, laid horizontally near the space center's entrance. Without the rocket, there'd be no indication that people monitored space launches from within the rectangular structures.

The Six all filed into a large conference room inside one of these boxy buildings—Building 1—with the twenty-nine male members of their astronaut class, and anxiously awaited their first assignments. This meeting, they'd soon learn, was the all-astronauts meeting, which took place every Monday morning. And since "all" astronauts were required to attend, that meant there were more than just thirty-five in the room that day. Another twenty-seven astronauts filled the conference area—a mixture of Apollo and Skylab fliers, as well as a good chunk of men who'd yet to make it to space. The two groups—the oldies and the newbies—sized each other up as they jockeyed for a place to sit.

With everyone seated or standing, the first thing people noticed were the wrinkles—or lack thereof. Most of the current astronauts were in their forties, pushing fifty, while many of the new were in their twenties and thirties. A palpable tension filled the air, both groups intimidated by the other, and everyone wondering how they should interact with the new faces in front of them.

Another divide cut between the military and civilian astronauts. Some of the military pilots and mission specialists who'd been selected already knew each other, after having served in the same squadrons and on the same aircraft carriers. But they were also bonded in the way they thought. Their military training had taught them similar behaviors, expectations, and humor, and learning to fly a new vehicle, even one as advanced as the Space Shuttle, would be relatively familiar territory. The women and male researchers, on the other hand, were from a range of backgrounds that included universities, hospitals, and the private sector. For them, the world of NASA was truly uncharted territory, and they had no clue what awaited them.

All these differences aside, one major factor set each person apart. Everyone understood what the presence of another astronaut in the room meant: competition. There was only so much room on each Space Shuttle flight, and every person translated to a butt that potentially could fill an empty seat. While the Six had a shared camaraderie from their first dance with the press, another thought tugged at them.

One of them had to be the first woman to fly.

The meeting began with the chief of the Astronaut Office, John Young, welcoming the recent selectees. He and George then explained what the new astronauts had to look forward to. The new recruits had a long two years ahead to prove they had what it took to fly to space. Technically, they weren't officially astronauts when they arrived at NASA on July 10. Instead, they'd come to NASA as astronaut *candidates*, or ASCANs. It was a new title and role—one never given to any of NASA's previous astronauts. From the Mercury Seven onward, all the men accepted into the program had been given the title of astronaut when they showed up on Day 1. But beginning with this class, years of training would be required before new recruits could be deemed ready to fly to space.

And that training was going to be diverse and grueling. The ASCANs would learn the ins and outs of the Space Shuttle's systems so that each man or woman could step in during a flight to fix a failing component. They'd tour NASA to better understand the various centers spread throughout the country. They'd learn an array of scientific and engineering disciplines, to better handle the satellites and experiments they'd be taking to space. And they'd fly in pairs tucked inside NASA's fleet of T-38 jets, to get a sense of operating a high-powered vehicle in the air.

It was a lot to take in, and after the meeting was over one of the new recruits stood up to make an announcement. The man was Rick Hauck, the most senior of the new class in terms of military rank. He'd also made the move of sitting at the table as soon as he walked in, which didn't go unnoticed. Rick passed around headshots for the new ASCANs to approve, a small administrative task but one that established him as something of a leader of this new cohort.

Once that task was complete, the newbies scattered to their first science lessons. NASA didn't waste any time getting started.

Before training got into full swing, though, the new group had some important business to attend to: They had to come up with a nickname—something snazzier than NASA Astronaut Group 8. Nicknames were a tradition that had begun with the first astronaut group, when they had received the title "the Mercury Seven." With thirty-five new astronauts in this class, the group had deemed themselves the Thirty-Five New Guys, or the TFNGs for short. Judy worked with Jim Buchli, who sketched out a logo for the class: a cartoon Space Shuttle with thirty-five astronauts crawling all over it. Members of the group would proudly wear T-shirts bearing the logo for years.

The seemingly harmless acronym actually carried an impish double meaning. The initials TFNG contained within them FNG— or "F**king New Guy," a reference in the military to a fresh-faced recruit who has to be shown the ropes. As the newbies of the astronaut corps, the TFNGs were winking at everyone.

The women appreciated being seen as inseparable from the larger group. "We didn't want to become 'the girl astronauts,' distinct and separate from the guys," said Kathy. "None of us had ever followed that model. I don't think any of us in our prior professional life had taken the path of being the girl x, the girl y."

NASA *had* to make some necessary changes to accommodate the women coming in, though. Many of the center's facilities hadn't taken women much into consideration when they'd been constructed. The biggest addition before the women's arrival had been a women's locker room attached to the gym, which included hair dryers and women-specific accoutrements.

Still, the Six were keen to blend in as much as possible, down to the outfits they wore. On one of the first days at work, Anna and Sally hatched a plan. The two snuck off to a nearby department store named Foley's, where they shopped for single-color polo T-shirts and khaki pants. The bland clothes were the unofficial uniform of the male astronauts and engineers. Going forward, it would be their unofficial uniform as well.

Dressing appropriately was only part of the process, though. Where the Six really had to prove themselves was in their training. Though Rhea, Judy, and Shannon had piloting experience, none had properly trained in a jet aircraft before. And in the event that a jet mission went sour, NASA wanted to ensure that the women and other novices could bail out of the cockpit—either over water or land—and survive. That meant the new recruits had to learn parachute survival training.

In late July, NASA flew the Six and the other male jet newbies out to Homestead Air Force Base in Florida, where the group would learn how to survive a bailout over open water. Over the course of three days, the fledgling astronauts suited up in a mix of dark green, orange, and light blue jumpsuits, and strode out through one-hundred-degree heat to nearby Biscayne Bay. There, waiting for them, were their parachuting instructors—and the press.

NASA's job was first and foremost to send people and spacecraft into orbit, but perhaps an equally important job for the agency was self-promotion. NASA's budget came from begrudging lawmakers, and, thus, NASA was constantly driven to showcase all the ways in which it was putting taxpayer money to good use. When a good photo opportunity presented itself, NASA didn't want to pass it up. And an event as juicy for photography as water survival training was the perfect way to showcase what NASA was up to.

A troop of reporters and television crews had bused into the area to witness and record the training. And, predictably, everyone wanted a piece of the Six. One reporter called out for Sally to give him a "happy look," to which she simply replied, "No!" Another tried to get Rhea's attention by yelling, "Hey, Miss." Rhea shot back, "It's Doctor." The male astronauts were again ignored, which prompted one to remark to a reporter, "We're mere commoners."

After offering up a few quotes to assuage the press, the women focused on the task at hand. The instructors buckled them into their harnesses, loaded them on boats, and shipped them out into the Atlantic. Soon, the Six were being dragged through the shark-filled waters, attached by rope to the back of a speeding motorboat. Their instructors challenged the women to keep their faces above water for as long as possible. Such a maneuver would be critical if they ever ejected and splashed down in choppy seas, with high winds pulling their parachutes to and fro.

The Six had to swim beneath floating parachutes, mimicking how to survive if the canopy came down on them during a fall. Once they emerged from under the chute on the other side, they'd inflate their life vests and wait for a helicopter to pluck them from the water. The finale of the training came in the form of parasailing—with a twist. With parachutes attached to their harnesses, the women sprinted off the deck of a boat, letting the wind sweep their chutes—and their bodies—four hundred feet into the air as another motorboat pulled them across the bay. When given the signal, they disconnected from the rope connecting them to the boat and drifted down to the water, role-playing a jet bailout. Once they hit the bay, they had to inflate a life raft and crawl on top.

For many of the women, this had been the most exciting and exhausting day of their professional lives. The barges filled with gawking reporters hadn't helped. "We're under enough stress doing things that we've never done before, especially the nonmilitary folks," Rhea recalled later. "Keep the press away from us." At one point during the training, Sally thought, *What am I doing here? I'm supposed to be a smart person.* But everyone came out of it all smiles. A photo from that day shows the women drying in the sun, grinning ear to ear.

Their parachuting trials weren't over yet, though. Ejecting from a jet over water was one thing. Coming down over land was another. NASA needed them to be prepared for any kind of evacuation. A few weeks later, the Six found themselves on another plane, this time to Vance Air Force Base in Enid, Oklahoma, not far from where Shannon had grown up.

On August 28, another excruciatingly hot day in Oklahoma, the Six and five of their male colleagues suited up for round two. This time, the women still had to sprint, with bulky parachutes hooked to their backs. But instead of racing across a boat deck, they charged over land as a pickup truck pulled them by a rope. Into the air they soared again, trying to re-create the experience of ejecting over solid earth. Once in the air, they released themselves from the cord connecting them to the truck and floated back down, positioning their legs just the right way to land safely.

Anna was one of the smallest in the class, and almost immediately the wind whipped her into the air. Conversely, the taller members of the group found themselves dragged by the truck for a while, waiting for a strong gust to get them airborne. Unlike after water training, the Six left land survival training battered and bruised. Judy messed up her ankle when she came in for a hard landing on the unyielding Oklahoma dirt. But the women held their tongues. As far as NASA knew, they'd loved every minute of it.

With survival training complete, it was finally time for the good stuff. NASA's fleet of sleek T-38s stood ready at the Ellington Field airport outside JSC, waiting for the Six to climb inside and streak into the air. White with a blue stripe down the side of its fuselage, each T-38 sported a thin conical nose and a curved glass cockpit, just large enough for a pilot and a backseater to fit inside. Twin General Electric J85 engines sat at the jet's base, capable of propelling the vehicle faster than the speed of sound and to an altitude ten thousand feet higher than a standard commercial airplane.

As part of ASCAN training, NASA wanted all the nonpilot mission specialists to get in at least fifteen hours per month of T-38 flight time. Since the women didn't have jet-piloting experience, they were officially back-seat riders, forbidden from taking off and landing the jets themselves. But even without that experience, they'd get to handle the aircraft in the air, work in tandem with their pilots, learn how to communicate through headsets, navigate through cloudy skies, and get slammed into their seats.

For the smaller women of the class, they soon realized their petite bodies were a challenge when it came to their flight equipment. Just stepping onto the ladder to get into the cockpit was a struggle for the five-two Rhea, and the training pilots had to search for chute harnesses small enough to fit the smallest of the Six, lest they slip out of their parachute during an ejection. But when it finally came time to fly, the women crammed themselves into the tiny back seat of the jets, strapped oxygen masks to their faces, and braced themselves as the T-38s practically leapt off the runway, taking to the skies over the Gulf Coast. Their first flights were all about familiarization, allowing the women to feel the plane accelerating through the air and hear the sounds of the cockpit over their headsets. But once those easy rides were done, the women would get their hands on the controls, temporarily assuming the role of pilot during a cruise. Sensitive to even the slightest movements, the T-38s required the lightest touch on the stick to bank or climb. "We fly by feel rather than deflection," said John Fabian. "So you feel your controls in your hands and you don't really move them a lot."

Each new flight challenged the Six to become ace jet pilots. During one flight, the women would find themselves being tested on their knowledge of instruments in the cockpit. On another, they'd have to develop a cross-country flight plan, mapping out where to stop for fuel. But perhaps the most exhilarating flights were the ones that turned the world upside down.

In the sky, the TFNGs would perform a nauseating array of aerial maneuvers, forty thousand feet above the vast Gulf of Mexico. The ASCANS would fly in tight formations, their T-38s soaring through the air together like a flock of birds, their wingtips mere feet apart. They'd execute backward loops, barrel rolls, and split Ss—a combination of a midair roll and somersault where the plane traced the bottom half of the letter *S*. If two T-38s flew in formation next to each other, one pilot could do a barrel roll and flip over the other plane—the two cockpits sweeping past each other as the twisting plane finished upright on the opposite side of its partner jet. If the pilots turned the aircraft just the right way, they could give their backseaters a heaping dose of extra g-forces, amplifying the pull of gravity and making it feel as if an invisible weight was slamming them into their seat. The TFNGs would get that same feeling again the day they strapped into the Shuttle cockpit and climbed into the sky.

It all could be stomach-lurching stuff. During her first flight, Rhea tried to remain calm as the pilot took her into a climb that made her feel weightless, trying to remember what she had for breakfast that morning in case she saw her meal a second time. But if anyone did get truly airsick, they kept it to themselves. Just as with their survival training, they didn't want anyone to think they couldn't hack it. "You realized that there was this added burden of wanting to make sure we succeeded so that we didn't at all affect the women who would come after us," Anna said.

Maintaining a pose of unflappability could be hard because things could get hairy during a T-38 flight. The jets didn't hold very much fuel, and in a few instances, some pilots barely made it home after an unexpected storm materialized on their flight path, causing their jet to gobble up more gas than usual. Once, Judy and her fellow TFNG pilot Dan Brandenstein found themselves landing in the middle of a freak snowstorm in Utah. Unable to see through the haze, Judy called out the altitude from the back as Dan came down on a runway she couldn't find.

"Five hundred feet," she said. "Four hundred feet. Three hundred feet. Two hundred feet." Then a pause. "Now what?" Judy asked. Legally, pilots aren't supposed to land if they can't see the runway at two hundred feet, and Judy tried to remain calm as they came down on what seemed like an invisible patch of earth. Fortunately, Dan was able to make out the lights of the runway just barely, and they touched down in one piece.

Shannon, who had thousands of hours of flight time by this point, took to the T-38 with ease, after adjusting to having "slightly" different controls from the Piper Clipper she was used to. The biggest disappointment for her was that NASA wouldn't allow her to get *fully* checked out in the T-38. Even the previous scientist astronauts NASA had hired in 1965—four of whom didn't have jet-piloting experience—went through air force training to command jets. But most of the TFNG mission specialists stayed firmly in the back seat of the T-38s, a requirement from NASA. Adding insult to injury, TFNG pilot Frederick Gregory, who'd flown mostly helicopters in Vietnam, had comparable fixed-wing experience to that of Shannon but still got to fly in the front of the T-38. Shannon asked higher-ups why she couldn't be in the front seat like Frederick.

"But Shannon, he flew in Vietnam and got shot at," one NASA official told her.

"Well, if I go out over the Gulf and get somebody to shoot some bullets at me, *then* can I get checked out?" she replied. Unfortunately, her request didn't work.

Being consigned to the back seat was particularly frustrating for Shannon because it meant being at the mercy of another pilot's schedule. The way it worked, the Six had to find a pilot who happened to be going somewhere outside Houston and hope their back seat was unoccupied. The Six quickly learned that not all pilots behaved the same in the cockpit, either. Some hogged the controls and didn't speak the entire flight, or they waged verbal battles with air traffic control, while others earned praise as "50 percenters," pilots who'd let the women have the stick for half of the flight.

Eventually, the Six found the pilots they loved flying with the most. For Rhea, that meant finding the best teachers and avoiding the thrill-seekers. Frederick Gregory had a penchant for flying as close to the ground as he could get, while another duo got in trouble for pulling off dive bombs on oil rigs out in the Gulf. A couple of pilots were also known to switch to abandoned frequencies over the Gulf and get into dogfights—terrifying aerial games of chicken that entailed two pilots flying at each other, waiting for the other to bail out of the dangerous tango first. Many considered it a miracle that no one suffered any fatal crashes those first few months of training.

Rhea found herself gravitating to the sandy-haired Robert "Hoot" Gibson, whom she'd introduced herself to at the astronaut presentation. For her it was about being comfortable, and Hoot, a TOPGUN pilot, knew a great deal about air traffic coming in and out of major airports. He'd prep her on what to say over the radio. "Air traffic control is going to ask you this, and this is what they want you to say," he'd advise her, helping her feel more confident communicating over the headset.

Judy, Sally, and Anna often found themselves in the back seats of jets piloted by Rick Hauck, John Fabian, Dan Brandenstein, and Jon McBride. Shannon often flew with John Creighton, who also happened to be her office mate. Technically, the women and other mission specialists weren't allowed to fly under five thousand feet. But a few pilots were known to look the other way, giving the women the opportunity to nudge the planes into the sky and take them down to the ground. "I let them land," John Fabian admitted.

Both Sally and Judy became highly skilled pilots. The men were amazed at how quickly they adapted to the controls, as if they'd been flying for years. "Judy was a natural pilot," John remembered. "Absolutely incredible ability in the air. And so was Sally." Hooked on this new pilot life, Sally even began taking private lessons on the side, eventually working to get her license. So did Anna, who eventually completed her first solo cross-country flight in a Cessna 150. That Christmas, she told Bill she wanted a book that one of her coworkers had on his desk, detailing all the airplanes in history.

And being a good flier went a long way at NASA. "The truth is that most of the skills you need to be a good Shuttle crew member, you learn in the jets," said Steve Hawley. "It's things like crew coordination. It's flight planning; it's talking on the radio; it's hand-eye coordination. If you're actually flying the jet, it's actually having to deal with real emergencies." The TFNGs began to realize that the better they performed in the T-38, the more attractive they became as potential crew members in space.

As exciting as the T-38s were, most of the TFNGs' time was spent in the classroom. For the graduate students and postdocs, that part felt a bit like home. The TFNG class was so big that they'd been separated into two groups, the Reds and the Blues, with Rick Hauck and John Fabian in charge. In the morning, the Reds, led by Rick, would get a briefing on a Space Shuttle system while the Blues, led by John, would fly in the T-38s. In the afternoon the teams would swap duties.

The classes covered every possible component of the Space Shuttle, from the complex hydraulics systems that used high-pressure fluids to manipulate all the mechanical flaps and valves, to the life-support systems that provided the critical mixture of Earth-like air in the cabin as well as water and waste removal. Despite their varying backgrounds, they all had to start thinking like engineers.

There was another important lesson to learn, though there was no official class for it: how to talk like a NASA engineer. Specifically, how to talk using acronyms. Practically every Space Shuttle system or component didn't have a name but, rather, a three-, four-, or five-letter abbreviation that engineers tucked away in the recesses of their brain. They might say: "On STS-51-F, SSME 2's hi-press FU T/P TDTS failed, causing early auto S/D of the center SSME at approximately three minutes prior to MECO." To the untrained listener, it sounded more like Klingon than English. But after listening to enough engineers, the Six learned to decode the acronyms and, eventually, use them freely themselves.

Other classes covered the various scientific specialties that would be represented in the Space Shuttle payloads and experiments. Much of the flight activity would involve studying the Earth from space, so the TFNGs learned about the planet's geography and oceanography, disciplines Kathy had mastered during her time at Dalhousie. There were classes on astronomy—a piece of cake for Sally—and lessons in human anatomy and medicine, in case a midflight medical emergency arose. For doctors Rhea and Anna, those classes were a breeze.

When they weren't in the classroom or the T-38 cockpit, the TFNGs spent the rest of their time traveling throughout the United States, visiting all that NASA had to offer. NASA isn't just a Houston-based operation. It included nearly two dozen centers spread throughout the country, a tactical move to disperse employees across different states and engender support from those employees' respective lawmakers. Each center had its own specialty, too. The TFNGs toured the Marshall Space Flight Center in Huntsville, Alabama, which oversaw the production of the Space Shuttle's crucial support systems: the massive external tank and the solid rocket boosters. The TFNGs also flew out to Kennedy Space Center in Cape Canaveral, home of the launchpads where the Space Shuttles would take flight—and the place that would become the astronauts' second home. And they visited the Jet Propulsion Laboratory in Pasadena, California, where engineers designed and crafted the robotic explorers that studied the solar system. During that visit, a few astronauts learned that not everyone at NASA supported the Space Shuttle program.

"I'll just never forget the director getting up and telling us that, 'We really [don't] need you guys,'" Hoot Gibson recalled.

Trips around the country were equal parts learning opportunities and PR visits. The astronauts would be shipped off to the headquarters of NASA's fleet of contractors, their mission being to show their smiling faces to boost company morale and, also, to glean a little more about the various vehicles that would help the Space Shuttle get off the ground. One such trip entailed a visit to Boeing, which at the time was working on the giant new 747 that was tasked with carrying the Space Shuttle on its back to and from

NASA's launch site in Cape Canaveral. George Abbey asked Rick Hauck if he, Dan Brandenstein, Dick Scobee, Judy, Sally, and Anna could all head up to Seattle to check out the plane.

The sextet paired up and flew in their T-38s from Houston to Seattle, to go behold what Boeing had to offer. They arrived at the company's factories and did their necessary meet and greet, before getting a tour of the production line. Then, during the tour, the Boeing handlers made an enticing offer.

"Do you want to fly the 747?"

The astronauts explained that they'd never flown an airplane that big in a simulator.

"No, we don't mean the simulator," the official replied. "We mean the airplane."

Boeing briefed the astronauts on the controls, and they loaded into the 747, a preproduction version devoid of passenger seats or other commercial service equipment. The massive fuselage stood empty, save for four seats. The six of them crowded inside the cockpit, as a Boeing test pilot took one of the seats at the controls. He then turned to the astronauts. "Who wants to make the first takeoff?" he asked. "Judy?"

The men hung back for a moment. Judy then looked around expectantly, before piping up. "Well, I would!"

She then took her seat next to the pilot. He taxied them out to the runway and revved the engines, while Judy had her hands on the controls, throttling up the plane. Judy then took the 747 out east over the Cascade mountains, completely at ease with the controls. After some time in the air, they returned to the runway. The Boeing pilot told the crew that he was going to demo a landing, and then everyone would get a chance to land the plane. All the astronauts exchanged glances, refusing to reveal just how much landing experience the women had.

Judy went first, plopping the massive plane down on the runway in a bumpy touchdown. She then took back off, and one by one, the other astronauts took their turns landing and taking off.

Finally, it was Sally's turn to take the joystick. As she came in for her landing, completely in control of the aircraft, the pilot asked, "So Sally, what other airplanes have you been checked out in?"

Nonchalantly, Sally shook her head and replied, "None," revealing she didn't have a pilot's license or experience flying such a massive plane.

The pilot's face went pale. He stammered, looking at Judy and Anna, silently asking with his eyes if they, too, had any experience. They both shook their heads sheepishly. The realization came over him that he'd just let three novices touch down the massive plane. But he couldn't have known—they'd all seemed like pros.

One plane that the Six became intimately familiar with they never got to fly themselves. It was a massive KC-135 that garnered the nickname "the Vomit Comet." Devoid of interior seats and filled with padding, the plane was normally used for aerial refueling but in this case was utilized to give its passengers a brief taste of the weightlessness of space. Through the sky, the Vomit Comet would fly in a series of parabolas, a path that would alternate the plane's inhabitants between brief periods of extra g-forces and periods of free fall. That way, the astronauts could get a feel for what it was like to float around the cabin of an enclosed vehicle—even if it's for just thirty seconds at a time.

The oscillations between weightiness and weightlessness often sent the contents of people's stomachs floating through the cabin, earning the plane its nickname. But the Six honed their lunch-retention abilities by enduring numerous peaks and valleys in the aircraft, sometimes up to fifty times per flight. They'd perform experiments and get a feel for what their Space Shuttle rides might be like. The Vomit Comet also provided the perfect test bed for NASA's latest invention: the Waste Collection System. Relieving oneself is fairly straightforward in Earth's gravity, but in weightlessness, one needs to be more creative. The engineers who designed the Space Shuttle's commode had created a very basic throne, with a four-inch hole that one could sit on while strapped to the seat. Also connected to the toilet was a hose, which sucked in air—and any other liquids that came out of the body. It was meant to be used by both men and women, but for it to work properly for the Six, a specialized funnel had to be attached to the end of the hose to make sure their urine would be directed the right way. Men can direct their pee; for women, it's not so simple.

Eager to know if the toilet design worked, NASA tasked some of the Six with testing the device on the Vomit Comet. Before each flight, they'd chug as much liquid as they could, filling their bladders to bursting. Then during their first stint of weightlessness they'd sneak to the toilet—located behind a privacy curtain—and attempt to relieve themselves in the hose. There was limited success—but plenty of toilet paper was on hand in case some drops escaped and floated away in the weightless environment. After the toilet testers had had their first go, it was time to chug water again. If they were lucky, they could give the throne another try before flight's end.

It may have been embarrassing, but it was better than testing the toilet on the ground. That test toilet had a camera inside, one that faced up through the opening to help astronauts find their perfect seat position. A monitor allowed them to watch their own anatomy as they made their . . . deposits.

But there was never any complaining, no matter how grisly or gross the training became. In everything they did, the Six knew they must project confidence, even when the pressures seemed overwhelming. One of those pressures had to do with working out and staying in shape, a daunting responsibility given their already hectic schedules. But the women didn't make any of their frustrations known. If a fellow ASCAN popped his or her head in the doorway and proposed a run in the 110-degree heat, you couldn't say no. The candidate would eagerly accept, all while praying for a cloud to come and block the sun.

The Six understood that as the first women, their every move was being scrutinized, more so than the performance of any of their male colleagues. They also realized that if one of them visibly messed up, critics would pounce, using the failure as evidence that women weren't fit for space. When Rhea found herself struggling with SCUBA training, she'd often break into tears as soon as she got home, fearing her struggles would be the end of her astronaut career. But she concealed her exasperation, showing up the next day, determined to press forward.

No one could see a hint of fear or weakness on their faces. But underlying everything they did was a silent acknowledgment of the enormous risk they were preparing themselves for. One training day, the ASCANs filed into a room to listen to a speech by Gene Kranz, the legendary flight director during the Apollo missions. He'd led the Mission Control team during the ill-fated Apollo 13 mission, when teams of NASA engineers helped save the crew from disaster when a fire started in their spacecraft's oxygen tank. The ASCANs listened intently as Kranz wrapped up his talk. Then he got up to turn on a recording that was piped through speakers.

At that moment, the ASCANs heard the voices of Gus Grissom, Ed White, and Roger B. Chaffee, Apollo 1's three crew members, as they sat on top of their rocket, undergoing a test. What the ASCANs heard was the trio's last moments as a fire broke out inside the all-oxygen environment and rapidly spread throughout the tiny, enclosed capsule.

"Hey! We've got a fire in the cockpit," Ed yelled over the speakers.

Roger's voice sounded, "We have a bad fire!" And then the men's final screams consumed the audio. After it was over, everyone in the room stayed silent.

It was a solemn reminder of what they risked in this job. And why their training was a matter of life and death.

CHAPTER 7

The Dawn of the Space Shuttle

The Space Shuttle was going to change everything.

That was NASA's dream for the successor to its moonshot when the program was cooked up in the early 1970s. This new 122-foot-long black-and-white spaceplane would represent a paradigm-changing transition from something dangerous and expensive to an endeavor that was cheap, routine, and safe.

The truth is, the Space Shuttle had started out as just a small piece of a much grander vision. As the Apollo program came to an end, President Richard Nixon searched for an answer to a critical question: What should come next for the US space program? To figure that out, he assembled the Space Task Group, spearheaded by his vice president, Spiro Agnew. They compiled a list of recommendations for human spaceflight programs that could serve as worthy successors to Apollo. The options spanned a wide range, including a wildly ambitious human mission to Mars, which would require more than doubling NASA's already extremely inflated budget. Along with this Red Planet mission, NASA could also establish a permanent base on the moon with orbiting outposts. Or the agency could defer the Mars mission and build a space station, which vis-itors would travel to via a "space shuttle"—a spaceplane that could make regular trips to and from space.

Fearing what it would cost to implement all or several of these options, Nixon eventually decided to focus on just one of the rec-ommendations: the shuttle. The space station would have to come later. After America won the Space Race, Congress's mood had turned frugal.

So, in 1972, with their instructions in hand, NASA officials got to work on the design of this human-operated shuttle. In essence, NASA envisioned the Space Shuttle to be America's most sophisti-cated truck—a space taxi that would conduct routine trips to and from low Earth orbit with the regularity of a semi that transported shipments and cargo across the country. And just like a truck, the Space Shuttle would be open for business, capable of carrying all sorts of payloads to orbit for a host of customers.

NASA's message to the country—and the world—was clear: If you need something to get to space, the Space Shuttle can get it there for you—even carry satellites to orbit (despite the fact that rockets without people on board had been getting satellites into space just fine for more than a decade). NASA hailed the Space Shuttle as the perfect platform for researchers to do science exper-iments in microgravity, as well as a great place to study the Earth from above. And NASA also sold the Space Shuttle as the ideal ride for the Department of Defense's super-secretive spy satellites; the astronauts on board would tend to those payloads, while keeping hush on what they were flying with. Plus, when NASA *did* want to build a space station, the Space Shuttle could carry modules into orbit, connecting them like LEGOs.

The vaunted versatility of the spaceplane ultimately dictated its final design. Central to the Shuttle was its payload bay, a sixty-by-fifteen-foot cavernous cargo hold that took up most of the vehicle's body. Once the spaceplane was free of Earth, two massive doors would open wide, exposing the bay's contents to the vacuum of space. If the astronauts needed to interact with the bay's contents, they could suit up in space suits to work with the payloads, or they could use a new tool being developed for the Shuttle: a dexterous robotic arm that could latch onto equipment and maneuver it through space.

With all these requirements, the vehicle morphed into a spacecraft completely dissimilar to the towering rockets-and-capsule combos NASA had launched years before. The Space Shuttle would *take off* vertically like rockets of the past, propelled by three main engines at its base. But with fixed wings on its sides, a vertical tail at its base, and a flight deck overlooking a bulbous nose cone, the Space Shuttle looked more like an airplane than a rocket.

And it would get to space with the aid of some important—and gargantuan—helpers. The Shuttle needed to gobble up half a million gallons of super-chilled propellants—the liquid versions of hydrogen and oxygen—to get its giant self into space. The amount of propellant needed was so unbelievably huge that NASA couldn't fit all the materials in the Shuttle. Instead, the volatile fluids would be kept inside a massive bullet-shaped burnt-orange chamber that sat attached to the Shuttle's underbelly, what became known as the external tank. During launch, the fuel would funnel out of the tank and into the Shuttle's main engines, where it would combust and rocket the spaceplane into orbit. When the propellants were all used up, the tank would be jettisoned and fall to Earth.

Still, the Shuttle's main engines alone couldn't provide *enough* thrust to get the spacecraft all the way to orbit. The pull of Earth's gravity was too great, and the Space Shuttle too heavy. So, on either side of the external tank sat two white rockets towering over 150 feet high. These were the solid rocket boosters, which did exactly what their name implies. They ran on solid propellants, a particle mixture of ammonium, aluminum, iron, and other combustible materials, to give the Space Shuttle the extra "boost" it needed to reach orbit. In fact, they packed such a big punch that the boosters provided the majority of the thrust during the ascent to space.

They'd become the largest solid rockets ever flown and the first ever used to propel humans to orbit—perhaps because no one had imagined before this that these types of rockets could be safe enough to carry such precious cargo. Unlike rockets that run on liquid propellants, a solid-propellant rocket cannot be turned off once it ignites. As soon as it starts burning, it burns through to completion; there's no stopping it. So, if something were to go wrong during a flight, the solid rocket boosters would just keep thrusting the craft forward, whether the humans on board wanted them to or not. For the astronauts, when the solid rockets ignited, they hoped they were pointed up.

If that didn't give the astronauts and mission managers enough to think about, the production process of a solid rocket booster was equally complicated. A company called Thiokol was selected as the main contractor for the rockets, and its engineers built the vehicles out of the company's factory in the middle-of-nowhere Utah. Because of this choice, Thiokol couldn't transport the rockets in one piece to the launch site in Florida; they were simply too big to make their way cross-country over land. They couldn't fit on a plane either. And in Utah there wasn't access to barges. Consequently, the company manufactured the boosters in segments, which were then shipped via train to Cape Canaveral. Each segment was as big as it could be while still fitting on a railcar. Engineers then assembled the segments in Florida, stacking them on top of each other with seams between each.

To make sure the piping-hot flames of the boosters didn't escape these seams, two thin rubber rings known as O-rings were placed inside each joint. These O-rings acted as seals, barriers to keep the scorching gases contained in the rocket. The O-rings were the biggest line of defense preventing the hot gases from escaping out the sides of the rocket and into the open.

In the end, a lot of moving parts had to precisely work together to get the Space Shuttle and its crew to orbit. The Shuttle's design could seem much more complicated than Apollo's, but the former had a major feature that NASA saw as a critical advantage: it came back to Earth intact. The astronauts, rather than return in tiny parachute-tethered capsules that splashed down in the ocean, would actually pilot the spaceplane, arranging for it to glide down to a runway. That made the system *partially reusable*, which was considered key to opening spaceflight up to the masses. Enormous amounts of money would be saved, as NASA saw it. By contrast, each Apollo mission had been like sailing the *Queen Mary* across the ocean *once* and then abandoning it once it reached shore.

True, the external tank was ultimately lost during each Shuttle flight, but parachutes sprouted from the two solid rocket boosters, allowing them to gently splash into the ocean where they could be retrieved, refilled, and refurbished. And the spaceplane that had returned was ready to be used all over again.

An early study from NASA and the Defense Department predicted a flight rate of between thirty and seventy flights a year, which could eventually lower the cost of getting one pound of cargo into space to between $50 to $100. That was the hope, anyway.

First, though, the newfangled spacecraft had to fly.

———

WHEN THE TFNGS came on board in 1978, they all believed that the Space Shuttle would be flying by the end of the following year. That was the target NASA still had in its sights. George Abbey had already picked two of the crew members for the inaugural mission: John Young, moonwalker on Apollo 16 and chief of the Astronaut Office, and rookie astronaut Bob Crippen, a handsome brown-haired former navy pilot with a gleaming-white smile and Hollywood tan. Bob had come to NASA in a group of eight in 1969 after the air force spaceflight program he'd been part of was canceled. John and Bob would be, respectively, the commander and pilot of STS-1—the first flight of the Space Transportation System, the Space Shuttle's more technical name. They'd be the world's bravest guinea pigs, flying on the new space shuttle *Columbia*.

But unbeknownst to NASA and its astronauts, STS-1 was still a few years away. And the TFNGs would soon learn that a NASA timeline is never to be fully trusted. It prompted Hoot Gibson to come up with a new definition for the term "NASA": Never A Straight Answer.

The Space Shuttle had been well into development by the time the TFNGs had arrived, but its road to the launchpad encountered several unexpected speed bumps. For one thing, the sophisticated new main engines that NASA had developed for the Shuttle kept blowing up during testing. Before flying the spaceplane, NASA would ignite the engines on test stands that held the hardware in place, to see if they performed as they were supposed to. Sometimes, the engines would produce a solid, uniform flame—a blinding-white cone of fire that quickly transitioned into a runaway plume of clouds that raced from the engine nozzle. Other times, that solid flame would start to flicker, bursting into a fireball that would engulf the engine in sparks and flames.

With each new explosion came a new set of problems to figure out. NASA pushed onward, tweaking the engines after each test and perfecting their design. But the engine tests stretched on and on, since NASA needed to ensure they wouldn't blow up on the back of the Space Shuttle during flight.

Simultaneously, another complication arose. The entire exterior of the Shuttle had to be covered in roughly thirty thousand insulating tiles made of silica fibers, fitted together on the outside of the vehicle like pieces of a puzzle. These tiles served as a crucial protective layer when the vehicle returned to Earth, preventing the Shuttle from burning up as the air surrounding the spaceplane heated to nearly three thousand degrees Fahrenheit. The installation challenge was that these tiles each had to be glued onto the Shuttle one by one, a process that could take up to forty hours for each tile. And some of the tiles just wouldn't stick. If the Shuttle got stressed during a test, the tiles would sometimes fall off or break.

It was because of such a problem that the first Space Shuttle arrived at its launch site half-naked. In early 1979, Boeing's 747 carried the sparkling-new *Columbia* on its back, transporting the spaceplane from its build site in California, where the vehicle's main contractor was located, to Florida for its first flight. One only had to look at the Shuttle in the sky to see that it was missing a large chunk of its tiles—roughly 7,800. Workers continued to install the pieces at Kennedy Space Center, but many of the tiles that had been glued on were already breaking and had to be replaced. Only after NASA found a way to strengthen the tiles, through a process called densification, did they stick for good and create a robust heat shield.

But the issues would push back *Columbia*'s first launch by two more years.

It was frustrating for NASA and government officials eager to see the Shuttle fly. But there was no room for error. Every part of the system had to be flight ready and safe, because the Space Shuttle lacked a key feature: an escape. Once the solid rocket boosters ignited, the spacecraft was committed to launching until the two giant candles burned out after two minutes. There was no way for the astronauts inside to bail out if something went wrong. The first Shuttle launches would be equipped with ejection seats (though fleeing a vehicle moving at hypersonic speeds wasn't a particularly desirable option), but the larger crews on follow-up missions would remain firmly bolted into their spacecraft.

Only after the solid rocket boosters separated from the vehicle could the crew potentially do something if a major system malfunctioned. If an engine went out on the way up, the Shuttle could potentially still reach orbit, albeit a lower one than planned, depending on when the failure occurred. There was also the possibility of the Shuttle's skipping orbit altogether and attempting a landing across the Atlantic in either Europe or Africa.

The most dangerous option of all was the dreaded "return to launch site," in which the Space Shuttle would attempt a flip maneuver in the air and head back to land on a runway at Kennedy Space Center.

Of course, all these aborts depended on the Space Shuttle remaining intact during the launch. If it broke apart during flight, the astronauts were helpless to change their fate.

Still, NASA needed the astronauts to be prepared for *any* kind of emergency that might arise. And that came with practice.

BY THE SPRING of 1979, the TFNGs had started to get a little more intimate with the vehicle they'd be flying. After months in the classroom, the ASCANs eagerly began getting actual hands-on experience with Shuttle equipment.

To properly train crews to fly on the Space Shuttle, NASA created two types of simulators—full-scale mockups of the vehicle's cockpit complete with all the seats, screens, and switches the astronauts would be interacting with during flight. One simulator stood still, while the other would move and shake, like a Disneyland theme park ride, re-creating the sensation of bursting into the sky during an actual flight.

In those early days, the TFNGs didn't get much, if any, simulator time. Priority was given to the prime crew astronauts, those who'd already been assigned to the first handful of flights. That meant John Young and Bob Crippen had the easiest access to the simulators. But the TFNGs got plenty of time to observe these lucky astronauts as they ran through all their checklists. And most gut-wrenching of all, they watched as the astronauts attempted to overcome every manner of malfunction NASA's engineers could cook up. Notorious NASA trainers known as Simulation Supervisors—or Sim Sups (pronounced like "soups")—would insert outrageous worst-case scenarios. Sometimes the astronauts juggled the catastrophes and safely landed the Shuttle back home or in the Seychelles. Other times their simulated vehicle would break apart into thousands of pieces, killing the entire crew. If that happened, it was time to start again.

While the TFNGs yearned for the days when *they'd* be sitting in the simulators as the next crew on deck, they filled most of their time working on individualized tasks. George Abbey gave each of the ASCANs their own engineering assignments to keep them busy as NASA ramped up toward the Shuttle's first flight. Each task was different, and almost immediately, the Six began to speculate on what the assignments meant for their future. In their minds, some jobs were good and others bad.

Sally managed to snag a juicy gig right off the bat. Her first assignment was to work with the Space Shuttle's fancy new Remote Manipulator System (RMS), or as it would become known, the robotic arm. Built by a Canadian company, the arm was a snakelike mechanical appendage that functioned a bit like a human arm. It had three joints—a shoulder, an elbow, and a wrist—except it had a bit more freedom of movement. It was so light that it couldn't carry its own weight on Earth, but in the microgravity environment of space, it was capable of lifting and moving satellites and payloads weighing thousands of pounds. The robotic arm would serve as the world's most sophisticated arcade claw game; astronauts on board would use the arm to pluck payloads out of the cargo bay and place them in orbit. Or the arm could be used to grab onto satellites already in space and place them inside the bay.

The arm wasn't a particularly easy thing to maneuver, though. Astronauts, stationed in the Shuttle's cockpit, had to manipulate the machine's controls while looking at video screens showing the arm's position in space. Translating what was on the screen into small muscle movements took impressive hand-eye coordination, something that Sally the tennis pro seemed to excel at. She'd come home from work and tell Bill that she thought she was pretty good at this robot arm thing—better than many of her colleagues.

Judy began studying all sorts of software—from the kind that would be used in the payloads to the type used in the robotic arm. She also worked with other astronauts like John Fabian and Mike Mullane to define the individual roles that mission specialists would have on board the Space Shuttle. But those assignments didn't last long. Eventually, she found herself assigned to the robotic arm, just like Sally. She took to it with ease, and it eventually became her specialty. Soon she was flying back and forth between Canada and Houston, helping to create procedures and develop software for operating the arm. It became clear to many in the astronaut corps that showing competency with the arm was a way to stand out.

Anna's initial role was also a big one: testing out space suits. NASA was in the midst of developing new suits for the Space Shuttle, and with the arrival of women—who tended to be smaller than their male colleagues—George Abbey wanted to see if the agency's engineers could develop an extra-small suit to accommodate more body types. During the Apollo era, each suit had been tailor made for the individual astronauts, but moving forward, the Space Shuttle suits would work a bit like a Mr. Potato Head. Astronauts would choose a torso section that fit them best, and then they'd attach the various legs and arms to complete the suit.

The problem was, NASA hadn't finished these new suits yet. So Anna had to wear an old Apollo suit worn by Pete Conrad, the shortest of the moonwalkers. Pete may have been small, but the suit still swallowed Anna, making the simplest of tasks a challenge. Space suits are really the world's smallest spaceships, just in the shape of a human body. They must house enough atmospheric pressure so that the person inside can survive—but manipulating the appendages of an inflated suit requires a lot of strength and dexterity. Having the right-sized suit is key to getting anything done. "For the smaller women if you can get a good suit fit, they can do just as well," said Anna. "But if you don't have a good suit fit, you're lost."

A couple days a week, she'd suit up in Pete's space suit, like a small child playing dress up in her big brother's clothes. A team of divers, photographers, surgeons, and more would help Anna lower down into a giant pool in the Weightless Environment Training Facility, or WETF, where she'd test out the suit. In the tank, she'd spy people taking her picture and waving at her, making her feel like a lab rat under surveillance. She'd soon learn this was just how NASA operated: an astronaut had better get used to people watching their every move.

Ultimately NASA would opt not to develop the extra-small space suit, which effectively preempted many of the women from being able to conduct spacewalks. "We're not discriminating. We're trying to be economical," a space-suit technician told a reporter. "You've got to draw the line somewhere."

Shannon found herself in a place called SAIL—the Shuttle Avionics Integration Laboratory—a separate facility that housed a working Space Shuttle cockpit, where astronauts could test out its software. Shannon loved it but others feared assignments like hers. The SAIL was located in Building 16, far from the Astronaut Office in Building 4. The ASCANs worried that an assignment that took them so far away from the rest of the team—and George Abbey's eye—would be detrimental to their chances of getting picked for a flight. It was something Kathy feared when she was assigned early on to test NASA's WB-57F aircraft, a vehicle designed to fly at extremely high altitudes, up to sixty thousand feet. On paper, it was an awesome gig. Kathy would get certified by the US Air Force to fly while wearing a pressure suit in the thin atmosphere—the first woman to get such a certification.

But her assignment took her out to Ellington Field Joint Reserve Base, more than five miles away from the Johnson Space Center campus. Kathy grew concerned that, so far from the astronauts' offices, she might be forgotten. She decided to make the most of it, though. The pressure suit she'd wear was the same one the crews on the first four Shuttle flights would wear. Perhaps, she thought, she could eventually leverage this skill into wearing an actual space suit one day.

While everyone else seemed to get a good assignment, Rhea wasn't thrilled with hers. She was assigned to help craft the food systems, a less than glamorous task when compared with manipulating robotics and testing software. "I ended up with the cooks," she'd say later. She'd hoped for something more technical as a way to learn more engineering, since her medical experience wasn't going to help much with the Shuttle's systems. A friend of Rhea's prompted her to question the assignment even more when they said they thought it was sexist.

Rhea tried to be optimistic though, convincing herself that she'd been picked because of her background in nutrition. Plus, no one argued with George's decrees. It just wasn't done. She threw herself into the assignment and was soon flying on the Vomit Comet with the new food packets the NASA engineers had designed, testing what would happen when they were squeezed in microgravity. Sometimes, the food packet tests happened concurrently with testing astronauts for motion sickness in a spinning chair. When the smell of fresh spaghetti and Parmesan cheese wafted over to the spinners, an up-chuck was likely to follow.

While everyone tried to rationalize their assignments, the truth was they were supposed to be confusing at first. "These technical assignments were deliberately intended—with malice aforethought—to put people where they would be uncomfortable," George said. "We put scientists in operational jobs, military test pilots in science, and so on. They had to be generalists, and they had to be able to cope with assignments that went beyond their experience."

The Six toiled away for months. They still had a year left in their training, and it seemed never-ending. But then, one day in August 1979, Chris Kraft showed up at the Astronaut Office with an announcement. The TFNGs were no longer astronaut candidates. Their two-year training period had been cut short. There was simply too much to do to prepare for the first flight of the Space Shuttle, and the new astronauts had progressed faster than expected. To commemorate their accomplishment, each of the TFNGs were awarded a small silver pin, a small totem that meant they were officially astronauts. But they all knew the title was meaningless. They'd only become astronauts when they flew—and they wouldn't fly until the Shuttle did.

———

ON DECEMBER 29, 1980, the completed space shuttle *Columbia* emerged into the Florida air, peeking out from the open doors of

NASA's gargantuan Vehicle Assembly Building in Kennedy Space Center under an overcast sky. One of the largest buildings in the world by volume, the white-and-gray VAB was once used to stack Saturn V rockets before they were moved to their launchpads.

Ever so sloth-like, *Columbia*, carried by a colossal gray platform the size of a baseball field, trekked outward on a river rock path that led to its launchpad: LC-39A. In just a few months, the spaceplane would be taking off from that exact site.

All the astronauts seemed especially charged in the days leading up to the flight. But no one had time to revel in their exhilaration. George had made sure that most of the astronauts had some kind of job associated with the flight, and the entire agency was slammed with last-minute preparations.

With their medical training, Rhea and Anna had been tapped for the search-and-rescue efforts. If the Shuttle had some kind of abort or crash during the flight, they'd be on hand to help save the crew. George put Rhea in charge of developing protocols for the astronaut doctors, after she came up with a plan for their training. She, Anna, and the others took a refresher course in trauma training, so that they knew how to go about saving someone's limbs—and life—amid a grisly accident. Then come launch day, the astronauts would work with elite, trained parajumpers who made up the search-and-rescue teams that would ride in helicopters and swoop down wearing parachutes to rescue the crew if they crash-landed back on Earth.

As head of the doctors, Rhea chose to be stationed at Cape Canaveral for the launch with NASA's head flight surgeon—the obvious location to see the Space Shuttle take off. She and the parajumpers would be ready if the Shuttle had to double back after takeoff and perform the "return to launch site" abort that everyone feared. They prayed their expertise wouldn't be needed.

Meanwhile, Anna would be waiting out in the bleached landscape of White Sands, New Mexico, where tiny gypsum crystals glistened like snow over the desert terrain. NASA had a landing strip out there, known as the Northrup Strip, which the Shuttle could use if it had to do an odd kind of maneuver known as the "abort once around." In this scenario, the Space Shuttle would still go to space and do a lap around Earth before coming back in for a landing. It was a fairly unlikely event, but Anna and her team of parajumpers would be on hand, ready to leap into action if they saw the Space Shuttle coming in over the white dunes.

Also stationed out in White Sands would be Shannon, who'd pulled chase duty. She and fellow TFNG Steve Nagel would also jump into play if the Shuttle did its strange loop around the world and came in for a landing at White Sands. The two would hop into a T-38 and chase the vehicle through the skies as it touched down on the strip, gathering as many visuals about the Shuttle as they could to report back to NASA. They knew there was a slim chance they'd actually fly along the Shuttle for this flight.

The days ticked by. Finally, the crucial month arrived: April 1981. After undergoing numerous tests on the launchpad, the space shuttle *Columbia* was ready to fly. In the days leading up to the flight, everyone assumed the positions dictated by their assignments. For Kathy and Judy, that meant flexing their performance skills. Both had been offered as experts to the major television networks, and they repeatedly studied their notes so that they'd quickly recall key Space Shuttle facts when asked by eager reporters.

Kathy had been assigned to ABC, but with Apollo 17 moonwalker

Gene Cernan already booked as the network's expert, she found herself providing commentary on ABC Radio. For her, the move was a blessing in disguise. She quickly realized that radio had its perks—mainly, that there was no need to dress up. Plus, the ABC radio booth was located three miles from the launchpad, providing a clear view of the controlled explosion that would send the Shuttle aloft. Kathy was happy to relay details about the Shuttle's various systems if it meant getting a front-row seat to a history-making flight and still being able to wear a comfortable outfit.

But at one point during the lead-up to the first launch attempt, Kathy overheard Gene give an incorrect explanation for how the Shuttle's computers worked. Wanting to convey the correct information, she tapped a nearby producer and asked if it mattered that Gene's answers were wrong.

That sent the producers into a tailspin, and over one of their headsets, Kathy heard the main TV producer shouting to get her on set.

"Next thing I knew, I was up on the TV set with Frank Reynolds—me happily dressed in radio casual," Kathy recalled later. "I was now on national television and facing the delicate challenge of contradicting the famous veteran astronaut sitting across from me on the news desk."

Judy was only a few yards away from Kathy at the NBC News setup, but she did have to play dress-up for the cameras. Sporting a bright pink blouse, Judy dutifully answered NBC anchor Tom Brokaw's questions during a segment on women astronauts in the days before the flight. She answered the basics about astronaut training, but then the subject of dating came up.

"What happens when you meet a man who's not in the space

program and doesn't know who you are, and you say, 'I'm an astronaut?'" Tom queried. "Does he say, 'Ah, you're too cute to be an astronaut. Come on little lady, you can't be an astronaut.'"

With a wide smile, Judy responded with the most diplomatic reply. "I just tell them I'm an engineer."

But he kept going, enamored with her personal life.

"What about the whole business about social relationships . . . Are some men threatened by the fact that you're an astronaut?"

"I don't know," said Judy. "If they are, they're probably not my friends."

"Are there discussions in Houston about what happens when men and women go into space for the first time, together?"

"There's not discussions among us," said Judy, adding later, "I think from our point of view, since we're so used to working together professionally, we look at each other as professional colleagues on the ground and in orbit and whatever. And we do it that way, period."

"Do you think the time will come when there will be romance in outer space?"

"Oh gee, I couldn't tell you that," said Judy.

It was a long couple of days.

The length was prolonged by the fact that *Columbia*'s first launch attempt on April 10 had to be scrubbed due to issues with the software on its computer. But two days later, on April 12, with crowds of spaceflight enthusiasts packed in tight on the beaches under Cape Canaveral's crystal-blue skies, the space shuttle *Columbia* was led through the final countdown by the flight controller.

"Seven. Six. Five. Four. We have gone for main engine start. We have main engine start. We have liftoff of America's first Space Shuttle and the Shuttle has cleared the tower!"

Sally watched in awe from the sky, looking out the back-seat window of a T-38 miles from the launchpad. Perhaps given the best assignment of all, she'd also drawn chase duty, assigned to follow the launch in the air. She watched as *Columbia* ascended, powered by its main engines and solid rocket boosters, looking like a small, inverted candle against a backdrop of vibrant blue. Rolling plumes of gas rippled out from the vehicle's engines, creating a puffy white mountain underneath *Columbia*'s base. The spaceplane then got smaller and smaller, until it disappeared completely from view.

High up in the sky, the two solid rocket boosters broke away and fell to Earth after two minutes of burning while *Columbia*'s main engines propelled the vehicle deeper into the darkness above. Over the next six and a half minutes, the vehicle ratcheted up to nearly 17,500 miles per hour—the ludicrous speed needed to reach orbit. Then at nine minutes after launch, a crucial moment: main engine cutoff, or MECO. The flames shooting from *Columbia*'s main engines disappeared while the Shuttle coasted through the vacuum of space. Seconds later, the external tank broke away, falling back to Earth to break apart and never be used again. During the final few seconds, smaller engines embedded in the Space Shuttle fired, giving the vehicle the extra oomph it needed to get to its final destination.

Eight and a half minutes after taking off, *Columbia* reached its orbit, putting it in a continuous loop around Earth. The era of the Space Shuttle had arrived.

CHAPTER 8

Working with Men

A little over a month after the space shuttle *Columbia* touched down at Edwards Air Force Base in California, Rhea Seddon donned a long white gown with transparent lace sleeves and walked into a small Methodist chapel in her hometown of Murfreesboro, Tennessee. Standing at the altar to greet her was a man with sandy blond hair and a thick mustache, wearing a black tuxedo. Hoot Gibson took Rhea's hands in his, and together the two exchanged vows to become husband and wife. When they exited the chapel, they were met by a crowd of a few dozen photographers. The camera bulbs popped and flashed as the press tried to snap pictures of the newlywed astronauts.

Hoot and Rhea had fallen in love in the sky. For Rhea, training in the T-38s had been a struggle at first, and she'd yearned for the right teacher to fly in her front seat. Hoot, over time, became that instructor. They grew close sitting front to back in the T-38 cockpit as Hoot guided Rhea and helped her become a confident flier.

They'd had their share of misadventures in the air, though. Once, en route to El Paso, Rhea took control of the plane after takeoff and almost immediately Hoot proclaimed "Dead pilot," meaning he was going to play dead. After a few curses, she took over as instructed. She'd try to ask a question, only to be met with "Dead pilot." That day, he'd tested her fondness for him.

On another flight out in California, the two had been practicing Shuttle chasing, pitching steeply up to thirty-five thousand feet as they'd have to do when the Shuttle for STS-2 was in front of them. All of a sudden, everything got real quiet—and Rhea noticed she couldn't breathe very well. Suddenly, she was hit by a terrifying realization: both of the T-38's engines were off. They'd flamed out during the complicated chase maneuver, which had stopped air from circulating through the cabin. With no engines working, they were adrift. Rhea tried to breathe calmly, while Hoot took the lead. He quickly shut off the throttles, stopping the flow of fuel to the engines. The two then waited in eerie silence as Hoot maneuvered the plane down to twenty-six thousand feet, where he could hopefully reignite the engines in the thicker air. If that didn't work, they'd have to perform a "deadstick" landing. Hoot knew of only one T-38 pilot who'd ever successfully pulled it off.

Fortunately, Hoot's quick thinking paid off. Once at the right altitude, the engines reignited and the two landed safely.

Though Hoot had been married at the time of his selection, his wife didn't take well to Houston. They broke up not long after he'd been selected and she moved back to California, splitting time with their young daughter, Julie. Rhea's heart had gone out to Hoot when she heard the news, and at the same time, her feelings for him grew. One day in Houston, the two had lunch together in one of JSC's cafeterias and Rhea came right out and said what she was thinking. "I feel myself very attracted to you." She'd made the first move again, just as she'd done with her handshake when the two had first met. Not long after, the pair began dating.

Later, Hoot would always joke that she'd been the one to pop the question, which many didn't doubt. Such a bold move fit her personality. But it was Hoot who finally asked. He originally planned to propose on Valentine's Day 1981, but the restaurant where he made reservations turned out to be so awful he decided to hold off. So, two days later, the pair went on a date to the nearby Kemah Boardwalk to redo their dining experience. After they ate, they walked out onto a pier and serendipitously spotted a shrimp boat floating by with the name *Rhea* plastered on its side.

At the end of the pier, Hoot turned to Rhea and said, "So are you going to marry me?"

Rhea cocked her head to one side and smirked. "Sure."

They became the first astronauts to marry while in the corps. And their marriage *would* have made them the first astronaut husband and wife, but Anna Fisher and her husband got there first. Bill Fisher never gave up on his dream of joining the astronaut corps, and when NASA initiated another astronaut selection in 1980, he applied and was accepted. Now, Anna no longer had to hold in her excitement about what she'd been up to at work when she came home. She and Bill could talk openly about what they were working on, with Anna providing some tips about ASCAN training.

Marriages aside, it hadn't been all smooth sailing between the men and women at first. When the Six first arrived at NASA, awash in a sea of men, they felt some initial unease. In 1975, just three years before the Six arrived, women made up about 17 percent of NASA's workforce, and the majority worked as technicians or in clerical roles. Many of the Six had grown accustomed to working in mostly all-male environments in previous jobs, but here they were especially outnumbered.

Exacerbating the "boys' club" feel was the fact that many of the male astronauts came from a hyper-red-blooded military background. They cracked sexist jokes with abandon, rarely considering whether the practice might offend their new female cohort. And they also reveled in their newfound celebrity status as astronauts. During trips across the country, a handful of men entertained the affections of women who were thrilled to meet space pioneers, sometimes missing the bus back home after engaging with fans.

This macho ethos permeated the office, too. When John Fabian first came to NASA, he'd hung a *Playboy* pinup calendar on the back of his office door. If someone walked in with the door swung open, the calendar would be hidden and out of sight. But if John and his officemate, Frederick Gregory, were inside with the door closed, the calendar was on full display. At one point, Judy discovered the hidden playmates and always made a point to pat the calendar on the way out the door.

Judy had done some decorating of her own to add balance. As a gift to the other five women, she handed out pink bumper stickers that read A WOMAN'S PLACE IS IN THE COCKPIT. The Six pasted them on their office doors with pride. But, perhaps predictably, their presence prompted abhorrent jokes. One day, when Mike Mullane and a navy TFNG he was with spotted the sticker, the latter turned and snickered, "A woman *is* a cock pit."

Women as aviators was just something that few men of the previous generation could wrap their heads around. And at times the Six found themselves struggling to be taken seriously by titans of spaceflight history. Apollo astronaut Alan Bean told *Texas Monthly* that initially many older astronauts felt that the incoming women were doing men's jobs. "I know I felt that," Alan said. "Astronauts, to me, were men; they had to think about computers and flying— male things." Once, out at Mojave Airport, Kathy had just landed in a T-38, ready to embark on a day of training to chase the Space Shuttle for an upcoming flight. While walking in the desert, she and her male colleague spotted Chuck Yeager, the famed test pilot who'd been the first to break the speed of sound. When Kathy's astronaut partner introduced her to Chuck and told him she was flying chase, the old test pilot scrunched his mouth and said, "Riding, maybe. Ain't flying."

It wasn't just the *astronauts* who came to NASA armed with bravado and a "men-only" mentality. Many of NASA's male engineers had to get used to working with women as their peers. During every major technical review and meeting, the agency would send an astronaut to represent the corps, with the Six in attendance on occasion. They'd be on hand to speak up on behalf of the astronauts, while also asking questions about things that didn't make sense. The engineers had become accustomed to the male astronauts talking to them as equals. But when one of the Six spoke up, a few of the men became flustered, not used to women questioning their decisions. Sometimes when the women attended, the engineers didn't realize that there even *was* an astronaut in the room.

The sometimes-chilly reception didn't just come from men. The Six also ran into friction from other women—not their female coworkers, but rather, the wives of their coworkers. During the first few weeks of training, a few of the pilots weren't too keen on flying with the Six; they blamed their wives, who weren't comfortable with women spending hours a few feet from their husbands. One astronaut expressed his frustration to Carolyn Huntoon. "I think it's great we're having women astronauts," he said. "It's my wife who doesn't think it's so good."

"That's not our problem," Carolyn said. "That's not the women's problem. That's your problem."

A few astronauts sometimes used perceived impropriety as an excuse to exclude one of the Six from her assigned role. During the third flight of the Space Shuttle, Kathy had been assigned to work at Cape Canaveral as a Cape Crusader, working the switch checklists for the flight. To accommodate the Crusaders, NASA had leased a three-bedroom condo at Kennedy Space Center. It was a super-convenient place to stay on the sprawling site, since it significantly cut down on the time involved in getting to the launchpad.

Kathy needed to head out early from Texas to Cape Canaveral, and she went looking for one of her fellow Crusaders, Don Williams, to get the key to the condo. When she found him and asked for it, he grew flustered.

"Well, actually, you know, I've been thinking about this," he said, as Kathy recalled. He told Kathy he was worried about the optics: two guys staying in the same house with a woman. "What will people say?"

Annoyed, she let him say his piece before responding as politely as she could. "I think how we handle it is entirely in our hands," she explained. "I think we just need to saddle up and go do it."

Don said he'd think on it. To help his deliberations, Kathy let the other women know what Don had said, and the five got to work. They all confronted him and let him have it. "What are you going to tell your wife when I'm assigned to a crew with you?" they'd ask. They reminded him that neither the media nor his wife would get a say in his flight assignments. It took Don a day before he finally gave up and handed over the key.

Moments like this had been anticipated, at least by Carolyn. She and other NASA officials were under no illusions that accepting women into the astronaut corps was universally beloved. But NASA had been determined to press forward, regardless of anyone's objections. "It's not like we were taking a vote," said Carolyn.

Carolyn established herself as something of a mother hen to the Six. If any needed help or advice on how to navigate JSC as a woman, she was ready to share her thoughts, as one of the few women to have a senior role at the center. And if there was ever an issue that the Six needed to address as a unit, Carolyn would hold informal meetings with them to hash things out.

One such issue was that the women had been plagued by royal blue flight suits that didn't really fit them. They'd been made for men, so the Six either wound up with a suit that was too tight around the hips or too broad in the shoulders. Eventually, NASA allowed the women to get them tailored, but the suits they had to wear in flight were just as bad and bulky. The pants came in small, medium, and large, and those were the only choices. "They were not very flattering," Rhea would later say.

One meeting overseen by Carolyn centered on the outsize attention the Six were receiving. The public affairs office at NASA would often accept appearance requests on their behalf, asking them to come talk to various groups or do interviews. And it was starting to weigh on them. Each appearance took a lot of work and effort, and sometimes involved taking off a couple days of work. That was valuable training time, and the days off were starting to add up.

Carolyn understood their frustration, but her advice was to

lower the volume. "By complaining about it all the time, they were really offending the guys who weren't being asked at all," Carolyn would later recall.

The Six felt that they could rely on one another when needed, and they were ready to pounce if they perceived sexism. But the truth was, they weren't exactly best friends. They were trusted coworkers, and they could form a united front when they needed—but outside of work, they didn't see each other much. Shannon was just too busy to socialize, with three young kids at home. "When I wasn't working, I was home taking care of my kids and doing all the things that you do with kids," said Shannon, adding, "Life was extremely full." Some of the Six grew closer than others. Rhea and Anna were both doctors, and both married to other astronauts. With those shared experiences, they naturally formed a stronger bond than with the other four. Sally, for her part, found herself liking Judy the best. The feeling was mutual. The two had similar senses of humor and were both fiercely driven. They were natural friends.

And some of the women felt as if they were on completely different life paths. "I was this girly girl from the South, and they were California girls," Rhea later recalled of Sally and Judy. "None of that [girly] stuff mattered [to them]."

In actuality, it was the *men* whom the women actually spent the most time with during their training. Since the Six were all backseaters in the T-38s, all their flying time was spent with a man in the cockpit. Those trips lasted hours, and the women and men chatted over their headsets, learning from each other and growing closer. Over time, the Six found themselves better friends with their male colleagues than with each other. And that meant putting up with their shenanigans.

The men exploited any opportunity they could to tease the Six, and Judy often found herself the butt of the joke. During a trip to Cape Canaveral with the rest of the TFNGs, someone put Judy's name on the door to one of the bedrooms in the crew quarters. When she went inside, Judy found pink satin sheets lining the bed. She "was not amused," Rhea remembered.

Another time while Judy was showering in the gym, a few of the guys found a grass snake outside and decided it would be a good idea to put the creature in her purse. They stood behind a nearby wall with childlike glee, waiting for her to open it. Sure enough, she exited the shower and reached inside her bag for a hairbrush, unaware of the purse's new occupant. When she put the bristles to her head, she couldn't manage to pull the brush through. That's when she looked down and saw the snake coiled up in the bristles. To the delight of the pranksters, Judy let out a piercing scream and the men promptly scattered to avoid facing their mark.

If you made a mistake on the job, you became a prime target for ribbing. Kathy, at one point, served as a Cape Crusader for STS-3, tasked with going over the thousands of switches in the Shuttle's cockpit just before midnight the day prior to flight. The process lasted hours, and there was no way for a sleepy astronaut to tag out when things ran long. After hours of struggling to keep her eyes open, Kathy was asked to respond to Launch Control over the radio. She pressed the button on top of the pilot's control stick to radio back, when suddenly all the buttons in the Shuttle's cockpit illuminated and alarms blared. Her sleep deprivation had caused her to mistake the button she needed to press. Normally, the button on the joystick in the T-38 is the one used to respond over the radio. The button Kathy had pressed switched the Shuttle's main computers into emergency backup mode.

The mistake threw Launch Control into a panic. Over the radio, the team spit out rapid-fire suggestions for possible malfunctions that had caused the computers to change settings. Mortified, Kathy eventually fessed up. Ultimately, the snafu was fixed, and the Shuttle launched just fine and on time. But Kathy knew she was in for it when she retreated from the cockpit that morning.

She walked into the post-launch party, braced for the ridicule. The launch director turned to greet her. In his hand was a gift: a gray metal box with two large buttons on it. One had been labeled "THIS ONE"; the other had been labeled "NOT THIS ONE."

Not to be outdone, Kathy solemnly announced that she'd already punished the culprit responsible for the accident. She held up her hand so everyone could see. The offending limb now had a large bandage wrapped around her thumb. It looked as if she had tried to sever the appendage. Everyone cracked up in unison.

The TFNGs didn't discriminate in their teasing, though. Everyone was an open target, men and women alike. Just as they'd hoped, the Six were close to attaining their goal: becoming "one of the guys." If the guys ever treated them like debutantes and not fellow astronauts, they'd shut it down. Hard. Sally, especially, made it crystal clear she wouldn't tolerate being treated differently. Once, Hoot watched in amusement as one of the male astronauts held the door open for her. Smiling, she shoved him through the door and held it open for herself.

Sally just wouldn't put up with that 1950s-era behavior. Judy was the same. One time, she accompanied John Fabian to the Air Force Academy in Colorado Springs, where he was set to give a lecture. He asked her if she could share some insights with the cadets along with him. When he walked onstage in front of the group, he introduced himself and then said, "This is one of our six lady astronauts who is in the Shuttle program." Judy walked out onstage and promptly corrected John in an aside, joking, "I'm no lady."

But despite a few hiccups in the beginning, overall, the women felt NASA was embracing them. They were part of the team, something that truly shocked Shannon, who came to the agency armed for battle against bigoted attitudes. "I was prepared for people to make all the typical comments that they've made all my life, and I was just shocked that it never happened," she said. "I mean, I do know that it was difficult for some people because [women as astronauts] was a new thing, but we were never treated in anything except a super-professional way."

Eventually the skeptical astronauts came around. "I changed that opinion," Alan Bean said of his initial attitude. He'd spent considerable time working with Rhea and taking on the role of her mentor, which may have played a part in changing his mind. "The job of astronaut is just as female as it is male."

Ultimately, the TFNGs became a family. It was impossible not to, given how much time they spent together. When they weren't at the Outpost Tavern, a shady dive bar near the JSC campus, astronauts took turns throwing parties at their homes each week. They'd also go on camping trips, as well as out-of-this-world excursions. Once, Hoot read about an upcoming total solar eclipse that would pass over the northwestern states in February 1979. He became determined to see it, but not from the ground. A poorly timed cloud could ruin the entire event. No, he wanted to see it above the weather, from the cockpit of a T-38.

He brought the idea up to some of the TFNGs, and soon dozens of astronauts wanted to go chase the eclipse in the jets. John Young got wind of it, and during one of the Monday astronaut meetings stopped the idea in its tracks. He told the group it would look unseemly if a bunch of NASA T-38s just showed up at Malmstrom Air Force Base in Montana. So he stipulated a condition: the astronauts had to have an astronomer fly in their back seat, or they couldn't do it.

Hoot got to work. He snagged Sally to ride in his back seat while Dick Scobee grabbed George "Pinky" Nelson, a mission specialist with a PhD in astronomy. As for Mike Coats, he partnered with Steve Hawley, whose background also included a PhD in astronomy. Together, on the day of the eclipse, the sextet flew up to Montana, racing out east in formation. As the moon crept between the Earth and the sun, the three planes sped along the path of totality, following the eclipse for as long as they could. Sally, Pinky, and Steve all arched their heads upward, snapping pictures as the dark orb of the moon blotted out the sun, casting an eerie incandescent haze across the sky.

With memorable moments like these, romances naturally bloomed. Steve had been intrigued by Sally from the beginning. When he'd been chosen for the Class of '78, a local reporter in Chile had interviewed him, asking why he wanted to be an astronaut. "It's your basic once-in-a-lifetime opportunity," he responded. Later, he read an interview with Sally in which she'd responded to a reporter's question the same way: it was your "basic once-in-a-lifetime opportunity." *Wow, she said the same thing I did*, Steve thought.

As Steve got to know Sally, he felt their similarities went deeper. They were both scientists focused on space. They were both avid

Dodgers fans. And as single TFNGs, they always seemed to find themselves in the same place at the same time. They played softball together two to three nights a week, attended barbecues, and hoisted a few at Friday happy hour.

But for a while Steve couldn't find a good opening to get close to Sally. When she first arrived at NASA, she was still living with Bill Colson. Steve didn't feel like he had an actual shot. But, months later, he heard through the grapevine that Sally's romance with Bill was losing steam. "I wasn't particularly fitting in," Bill confessed. While Sally threw herself into her work, Bill had been doing research at nearby Rice University. Sally had wanted Bill to try becoming an astronaut, and he'd even taken flying lessons with Bill Fisher, Anna's husband. But after many weeks of trying to fall in love with space exploration, he realized his heart just wasn't in it. "I . . . found it a little boring." To him, the life of an astronaut meant following rigorous procedures and making sure nothing surprising actually *happened*. A good astronaut had to be obsessed with safety first, and that just wasn't him.

Eventually, Sally initiated the split. Less than a year after they'd moved to Houston together, she told Bill she wanted to live alone. She didn't say she wanted to break up, but Bill got the message. Looking back, he figured she *did* want to stay together. But that was just Sally's matter-of-fact way of handling things. She'd wanted more freedom, and that was her way of getting it.

Bill moved into the dorms at Rice, and though they kept in touch as friends, Sally became a free agent, exploring her options. She briefly dated Hoot at first, just for a few months before he realized that Rhea was interested in him. Again, Steve figured his chances were basically over.

Then in 1979, Hoot and Sally's relationship fizzled. Soon after, a group of TFNGs rented a couple of cabins in Texas Hill Country. They planned to go floating down the Guadalupe River in inner tubes, while imbibing adult beverages. Steve hadn't wanted to go, but the night before everyone was set to leave, Sally asked Steve if he'd come along. Her request ignited a small spark. "That was the time when I thought that, you know, well, I don't know, maybe she's interested in me," Steve said. "And I'm interested in her." It gave him enough courage to pursue Sally. Not long after, the pair became an item.

Meanwhile, Judy was embracing the freedom that came with being a single astronaut. She had a reputation for being able to hang out with men on their own terms. Sally was known around the space center as the activist; she made it clear she wouldn't put up with any sexist jokes or caddish behavior. But Judy would just smirk and laugh off a dumb joke—or shout an expletive in response. It was the reason the TFNGs felt so comfortable pranking her. She'd made it clear she could take whatever was thrown at her and throw it right back. "Judy had that unique way of navigating between being one of the guys and being a raging feminist," Mike Mullane said. "She'd give as good as she got."

Throughout her training, Judy had kept in touch with Len. The two would still see each other from time to time. He'd come to Texas, or she'd hop on a plane and visit him in some other city. But they were never fully together during that time. Judy was having too much fun in her new life. She was fast becoming friends with her TFNG coworkers. Thanks to her work on the robotic arm, she grew close with John Fabian as they traveled to and from Canada to work on the system together. One night after a long day of work on the arm, the two went out for beers and smoked cigars. "Judy was my best friend in the Astronaut Office," said John. "I thought the world of her."

Life was nothing but sweet for the Six and their fellow TFNGs. Together they were having the time of their lives, forging bonds they knew would last for decades. And all of them commiserated over their shared anxious fate. Each TFNG yearned for the day that they'd get the call from George, telling them they'd been assigned to an upcoming flight.

CHAPTER 9

Choosing "The One"

On August 21, 1981, George Abbey was standing in his office on the eighth floor of Building 1 when he heard a noise outside his window. He turned to see a man dangling from a rope, dressed in overalls. It wasn't just any man, though. The stranger unbuttoned the straps of his overalls to reveal a Superman costume. The Man of Steel then pounded on the glass while singing "Happy Birthday" to George, who that day was turning forty-nine. As soon as the dangling crooner finished the song, he rappelled to the ground and disappeared from sight. George just stood there and blinked.

NASA would eventually identify Superman's secret identity. It wasn't Clark Kent, but a NASA astronaut by the name of Jim Bagian. Selected as part of the 1980 astronaut class, Jim had conspired with another 1980 astronaut, Guy Gardner, to get some of George's highly desired attention. The two had persuaded JSC security to give them access to the windows on the ninth floor, above George's office, claiming they were window washers. Jim had then rappelled down to his boss's window to perform the stunt.

The window concert didn't go over well with security or NASA management. Jim and Guy eventually got a talking to from John Young, though it amounted to no more than a slap on the wrist.

Despite the admonishment, the stunt *had* achieved what Jim had wanted: George's notice. For many of the astronauts, George Abbey's attention was prime capital. From Day One, it became immediately clear that this man ran the show. He picked the astronauts; he gave them their technical assignments; he assigned them to crews. The fate of an astronaut's career rested in the palm of this man's hand. And George's ever-watchful eye was always on the spacefarers, even until the moment they blasted into orbit. For each launch of the Space Shuttle, George was in Cape Canaveral, walking the astronauts out to the vehicle that took them to the launchpad. Given his near-total control over their professional lives, everyone hoped to get on George's good side, surmising that achieving his favor would eventually lead to a crew assignment—and hopefully a good one.

"One of our standard adages in the office was there's no such thing as a bad spaceflight," said Kathy. "But flying sooner is better than later. A longer flight's better than a shorter flight. High inclination is better than low inclination"—referring to the route the Space Shuttle took around Earth—"and spacewalks are wonderful." But astronauts weren't exactly picky, either. "When in doubt, refer to rule number one: there's no such thing as a bad spaceflight," according to Kathy.

But the only way to get a spaceflight was to get a call from George.

The astronauts' frustration lay in the fact that there was just no way to predict what kind of assignments any of them would get and when. To the astronaut corps, George's decision-making process was a mystery—as was George himself. A large man with a crew cut and droopy eyes, he wasn't given to many words, and the words he did choose usually came out in the form of a low baritone mumble. His crew selections always occurred in private, done only in consultation with John Young or a few other high-ranking officials. And when a new crew was announced, he didn't offer detailed explanations for his choices. The astronauts simply accepted the selections while secretly wondering about them in hushed tones over beers and cocktails after work. "That was the biggest hobby in the Astronaut Office, guessing crew assignments," said Dan Brandenstein.

Thus, George became an almost mythical figure at Johnson Space Center. To some of his recruits, he wasn't just George but King George. A few astronauts theorized that being at the top of George's mind was what it took to get selected. Each week, George would hold court at either the Outpost Tavern or Pe-Te's Cajun BBQ House during Friday happy hour, with many of the astronauts attending simply to get face time with the man who could put them in space. Some came to love the quiet boss and his mild-mannered ways, while others grew frustrated over his lack of rationale. Those who became George's drinking buddies were given the nickname FOG: friend of George.

Over time, George's mythos grew, culminating in more than a few practical jokes. During STS-5, when the crew walked out of their quarters and made their way to the Shuttle, George was by their side, as was tradition. But one of the astronauts carried a small white sign. It read "NASA OFFICIAL GEORGE ABBEY," and next to the words was an arrow. The astronaut pointed the sign at George as he walked. The crew giggled to themselves as they strode to the van that would take them to the launchpad.

At one point, the pranks even extended into space. While

working together ahead of STS-2, Steve Hawley and Ellison Onizuka were in the cockpit of *Columbia* in Florida, going over the switches before the crew came on board for flight. At one point, Ellison pulled out a surprise he'd snuck in. It was a picture of George attached to a piece of hardboard, with a decal that read: "GEORGE W. S. ABBEY, OUR LEADER AND INSPIRATION." Using Velcro attached to the back of the board, Ellison stuck the decal on the wall of the cockpit just before takeoff, a small token to amuse the crew. As a result, George's visage made it to orbit.

The photo wasn't meant to be seen. But when the crew returned, NASA processed the photographs that had been taken on board. Sure enough, one of the astronauts had accidentally snapped a shot of George's portrait in the background. After he saw the photo, George summoned Steve and Ellison to his office. George, with the slightest smirk on his face, pointed at one of the men and said "Wheelus." He pointed at the other and said "Incirlik." The two words were names of air bases in Libya and Turkey. George was giving them the locations of their next "assignments."

Ultimately, George could appreciate a good gag, but for the most part, he ran an incredibly tight ship and rarely strayed from the rule book. Over time, the astronauts came to realize that he rewarded those who did their work and worked hard. But there really was no telling if insubordination helped or hurt. "One of the complaints that you heard from the astronauts was you never knew where you stood," said Hoot Gibson. "He wasn't the kind of person who'd tell you if you messed up." Jim Bagian would ultimately get assigned to two flights on the Shuttle, despite his Superman stunt (albeit about eight years later). And those who *did* get assigned to flights early were often just as in the dark about their selections as everyone else.

George, for his part, never considered his choices that opaque. His secret sauce wasn't so secret. For every Shuttle flight, he simply tried to pick the people with the best skills who could operate the payloads and experiments on board. "There wasn't any real magic about it," George said. "We tried to match the requirements for each mission to those who were qualified and could meet those requirements."

And that's exactly what George tried to do when he began assigning the TFNGs to their first flights. By 1982, crews had been assigned through STS-6, and the time had come to select the next three missions: STS-7, STS-8, and STS-9. Each mission had its own unique requirements, which would ultimately guide George's selections. But he also had another plan in mind.

America was about to see its first Black astronaut fly. And its first woman.

NO ONE TALKED about it openly, but the question was a steady drumbeat underlying everyone's thoughts. Which of the Six would become The One? Sure, technically she wouldn't be the first woman to *ever* fly in space, but she'd be the first *American* woman, and that would still hold tremendous weight. Her name would live on the elite list that included all the major US spaceflight pioneers, including John Glenn and Neil Armstrong. She'd be a pathbreaker whom history books would reference for centuries. Instantly, she'd become a hero to millions of young women around the country. And becoming The One would bring with it celebrity, contracts, validation, speaking fees, and a larger-than-life identity. The other women would still do groundbreaking work, but their names wouldn't resound as loudly or as long.

The silent competition began almost as soon as the astronaut class was formed, playing out first in newspapers and magazines and on television broadcasts. The press corps quickly categorized the Six into various archetypes, showing preferential treatment to those who fit the more feminine, idealized version of womanhood. Anna, for instance, had become something of a media darling. Before the women even showed up for training, *Redbook* featured her on its cover, along with a detailed article discussing her exercise and diet routines (despite her routine at the time being "eat dinner and sleep" after twenty-four-hour shifts at the hospital). With her award-winning smile, the slim, feather-haired Anna exuded the classic charm that the press loved, and she eagerly sat for interviews when asked. Her face appeared in numerous magazine features, and she could often be found espousing the merits of the Space Shuttle program on TV.

Anna's willingness to perform for the press wasn't a calculated move to put her at the front of the pack. Her motivation was simply to do whatever it took to keep the Shuttle program afloat. "I just felt like, if there's anything I can do to get good publicity and to get this Space Shuttle off the ground so that I have the career I want to have, I'll do it," said Anna. "If the Shuttle didn't fly and things weren't successful, I was going to be figuring out what I was going to be doing."

But over time, it seemed as if Anna had become the face of the astronaut corps. And when Bill was selected as an astronaut in 1980, the press's fascination with Anna and her husband only grew. They became America's first astronaut couple, catnip for the ravenous media. "A MARRIAGE THAT WAS MADE FOR THE HEAVENS," the *New York Times* declared. Together, they seemed like a model celebrity couple. "The Fishers look as if they had been chosen by central casting rather than NASA," the paper stated.

The cute blond surgeon Rhea wasn't far out of the press's sight, either. Pictures of her submerged in the Gulf of Mexico during water survival training or floating weightless in the Vomit Comet, a giant grin spread across her face, dominated newspapers and magazines. *Weight Watchers* featured her in their magazine, touting "THE LADY ASTRONAUT'S DIET." "She looks like a college cheerleader—one of the nicest and prettiest girls on sorority row," the article read. And when she and Hoot married, the press also delighted in *their* love story. Now, the country had *two* astronaut couples, a concept that seemed so foreign just a few years prior.

Suddenly, everyone wanted to know the same thing: Would the couples eventually fly into space together?

The other members of the Six saw it all playing out before their eyes. And they knew where they stood. "If you were going to line the six of us up, put our six photos up, and pick cover girl shots, I'm not a cover girl type," said Kathy. "I've been on covers, but I'm not an archetypal cover girl look or face; neither is Shannon." Sure enough, the media liked to emphasize Kathy's weight and height, noting her larger size compared to her more petite colleagues. Shannon was the tallest and already a mom of three when she came to NASA. She was quickly cast as the mother figure of the bunch, a little less exciting than her single or newlywed colleagues.

Both Sally and Judy could have also been considered "cover girl types," but they had little interest in placating the press. Sally's attempts to keep the media at a distance were a matter of shyness. Her introverted self was never one to seek the limelight. She'd also gotten an early taste of the press when she first started at NASA, and she didn't like the asinine questions the mostly male reporters asked. Instead, she preferred talking to one female reporter, Lynn Sherr, a correspondent at ABC who focused more on the missions and training. As for Judy, her feeling was that her life was none of the press's business. She didn't like the focus the media put on her relationships—notably her divorce—and she wasn't interested in giving the public a guided tour of her complicated upbringing. She rarely gave interviews, and when she did her answers were succinct—no embellishment. As a result, America heard very little about Sally and Judy in the program's first years.

NASA wasn't openly grading anyone on their star quality, but everyone figured that an astronaut's PR value would probably be part of the selection equation. The One would certainly be inundated with speaking requests and TV appearances. Weighing her chances at the time, Kathy recalled, "I've got to bet there's some big factor that, all other things being equal, they will pick a cute one who gets on lots of covers. This is all guys making this choice. Somehow that's in there. So that probably means Anna, Judy, Rhea, and Sally have a qualitatively better chance of getting the nod first than me or Shannon."

Shannon could feel it too: the politics at play underlying the decision. For her, being The One wasn't really what she strived for. She just wanted to work and do a good job. Most of all, like any astronaut in the corps, her overarching ambition was to fly. And according to Kathy's rule, flying sooner is always better than later. "Being first was one thing, but you want to fly as soon as you can," Shannon said. "Waiting is hard."

While the American press corps picked its favorites, the astronaut corps also placed imaginary bets on which woman would be picked first. To them, the evidence was visible in what the women were doing, not the number of newspapers they appeared in. That being the case, everyone kept a watchful eye on the kinds of assignments the Six received from George. And despite their aversion to the spotlight, Sally and Judy seemed to be getting the best tasks. They'd become masters of the robotic arm, which was turning into a critical system for the mission specialists to know. The dexterous robot had so much versatility in what it could do: it could grab things, including satellites and suited astronauts; it could dispose of objects in space; and it could take pictures and video with a camera embedded at its tip.

Everyone felt that it was either going to be Sally or Judy, with Anna in the mix. Along with her space-suit work, Anna had also started work on the robotic arm and had become just as adept at the controls. But one major move pushed Sally ahead of the pack. For STS-2 and STS-3, she held a coveted role as Capsule Communicator, or CAPCOM. For every flight, CAPCOMs would relay crucial communication over the radio to the astronauts in space, serving as a liaison between the crew and the dozens of people in Mission Control. It also gave Sally a better understanding of what the communication between the astronauts in orbit and those on the ground would look like. The position had put Sally in the center of the action, JSC's Mission Control Center, for the second and third flights of the Space Shuttle. And she became the first of the Six to play this role.

When Sally moved to CAPCOM duties, Judy took on a leadership role with the robotic arm, masterfully manipulating the controls as if the appendage was just another arm on her body. And though she didn't get assigned as a CAPCOM, she did get to spend time in Mission Control too for STS-3, as the go-to robotic arm expert. That flight involved rigorous testing of the arm in space, and Judy was asked to be on hand to confer with flight controllers when her expertise was needed. When it came to robotic arm mastery, Sally and Judy were neck and neck.

In the end, though, there was just no way to tell who was going to get the nod. "It's kind of difficult to fight tooth and nail when you don't know what the rules of the game are," Rhea said. "About the only thing you can do is work as hard and do as well as you can."

By early 1982, no announcement had been made yet. It was still anyone's game. And while the astronaut corps continued to wait for this fateful decision, one of the Six harbored a secret she knew could take her out of the running.

Under her clothes and her flight suits, Rhea was sporting the smallest baby bump.

She'd found out she was pregnant right after the second flight of the Space Shuttle in late 1981. During a routine OB/GYN appointment, her doctor had noticed Rhea's particularly large uterus and opted to do a pregnancy test on the spot. Rhea was shocked when the results came back positive.

That same day, Rhea searched for Hoot in the SAIL, where he was working. When she found him, she asked him to meet her in the lobby and then slipped some pictures into his hands.

"Do you have some time to look at some mystery pictures, dear?"

Confused, Hoot tried to guess what he was looking at in the grainy black-and-white photos. He thought the pictures were another TFNG prank, maybe the ghostly outlines of someone's butt from a Xerox machine. Then Rhea pointed out attributes in the picture: a skull, a heart, and a little body. Realizing it was a baby, Hoot asked whose it was.

"Ours," Rhea replied. Hoot gasped, thrilled—and surprised that the pregnancy happened so soon after their nuptials. The two celebrated privately, then came to an agreement. They wouldn't tell NASA right away. Rhea feared management would ban her from parts of her training, notably flying in a T-38. Combining the caution of a medical doctor and the undercover skills of a spy, she began gathering as much research as she could to show that she could still safely fly in a high-performance jet while pregnant. When the time came, she wanted to present NASA with a portfolio of evidence: *Here, I'm pregnant, and I can still fly.*

There was one problem though. There was a near-complete lack of data on how flying in a T-38 might affect her pregnancy. The biggest dilemma was the possibility of ejecting and whether that might put the baby at risk. But it was another NASA blind spot. No one at NASA had ejected from the T-38s in many years, and certainly none had been pregnant.

So Rhea kept quiet. She climbed into the T-38 cockpit for months, not saying a word to anyone about her secret. But finally, in early March 1982, she was forced to come clean. She'd found out she was going to have a little boy, and her growing bump was getting too big to easily explain away.

One day, she and Hoot formed a united front. They drove to work and proceeded to go on a tour of upper management's offices. One by one, they spoke with John Young, George Abbey, and then Chris Kraft. Rhea made it clear to each that she didn't plan to quit working. She still wanted to continue her job in the avionics lab, where she had recently been assigned, and she very much intended to stay on as a helicopter doctor for the upcoming STS-3 mission.

John Young didn't quite know what to say, while George mumbled some short reply. Chris Kraft, however, had sensed something was up. He seemed pleased that Rhea was showing commitment to her career.

Rhea returned to her office, content with how the meetings had gone. Nobody had freaked out. Nobody had said no. It was probably the best she could have hoped for.

Then, as soon as she sat down, her phone rang. On the other end was one of JSC's flight surgeons. He told her she was no longer allowed to fly in the T-38 while she was pregnant.

Shocked, Rhea tried to explain that she'd gathered data on this and that she felt she'd be all right. She started going through the data, but the surgeon didn't want to hear it. NASA didn't want to risk the bad press if something went wrong with her pregnancy during a flight. Rhea then wanted to know: How was she supposed to travel to California for STS-3 to perform as a helicopter doctor if she couldn't fly in the T-38?

"Fly commercial," the surgeon said. Case closed. Then the call ended.

Rhea placed the phone in its cradle, feeling dizzy. This was exactly the kind of response she'd dreaded. NASA was still run by men—men of a much older generation—and their conception of what pregnant women could handle was outdated. But Rhea knew there wasn't much she could do. Fighting back was out of the question lest she be seen as "difficult." She held her head up and got back to work.

At the next Monday all-astronauts meeting, Hoot broke the

news to the rest of the corps. John Young asked if anyone had any last comments to make. Hoot raised his hand and stood up. "I'm here to announce that our total number of astronauts has increased to eighty . . . and a half," he said. The room cheered, with everyone ecstatic for their colleagues. After being banned from the T-38s, the warmth of the response from her fellow TFNGs touched Rhea's heart.

A month later, however, George was ready to make a decision. He'd been mulling over the assignments for the seventh, eighth, and ninth Shuttle flights, and he'd made a final list for each crew. These would be the flights that would take up the first TFNGs—and they'd be the ones who made history.

The seventh flight, STS-7, would be unique. While the flight would deploy two telecommunications satellites for Canada and Indonesia, it was also going to carry an experimental payload known as the Shuttle Pallet Satellite, or SPAS-01. NASA planned to use this payload to test out something the Shuttle had never done before, a type of maneuver known as a rendezvous and proximity operation (RPO). Essentially, the Shuttle would be performing a graceful tango with the object while in space. The crew of STS-7 would deploy SPAS-01, and the pilot would then fly the spaceplane close to the satellite, hovering near it, maintaining a steady distance. Then, at some point, they'd attempt to grab hold of SPAS-01 while it was flying free. It was going to be a very delicate and complex operation that required great precision—to ensure that the Shuttle didn't accidentally collide with SPAS-01 while trying to grab hold of it.

To get SPAS-01 into space, STS-7 would need highly skilled robotic arm operators. Not only would they use the arm to lift the satellite out of the Shuttle's cargo bay but they'd also use the arm to pluck SPAS-01 out of space and tuck it back inside the vehicle at the end of the demonstration.

George already had a commander in mind for this flight: Bob Crippen, one of the stars of STS-1. Bob, also known by his nickname "Crip," had done a terrific job during the Shuttle's inaugural mission, and George felt he was ready for a leadership role. For the pilot position, he went with Rick Hauck, the unofficial leader of the TFNGs. After that, it came down to the mission specialists. George wanted John Fabian, who'd been lead on the robotic arm at one point.

Both Sally and Judy had worked closely with John. He'd given them both glowing reviews. They were both fully capable. But only one would go up. To partner with John on the arm, George decided he wanted Sally.

George presented his crew ideas to the space center director, Chris Kraft, along with the crews for STS-8 and STS-9. For STS-8, another mission that would rely heavily on the robotic arm, George had assigned Guion Bluford as a mission specialist. That would make him the first Black American to fly to space. Back-to-back, the two flights would be groundbreaking for the space program—and for America.

But when Chris saw the list, George received a surprise. Sally Ride was a no. He wasn't convinced by George's pick for the first woman. "Why not Anna?" he asked. Chris felt that there were at least two women above Sally who were qualified.

George was in a bind. He really believed Sally was right for the job. Even though he was regarded as the "astronaut maker," he often had to sell each choice. As he'd later say, "I had to justify to Kraft why I picked each and every one of the individuals I picked."

George went back to his office and got to work laying out the reasons for his decision, getting input from key players. One day, he asked Bob Crippen to come meet in his office. When Bob entered, George, in his signature matter-of-fact style, asked the STS-1 pilot if he'd like to command STS-7. Jubilant, Bob told George he was more than happy to do it. Then, George broached the topic of who else might be on the crew, specifically Sally.

"We all knew that whoever was going to be the first woman to fly was going to get more attention than they'd probably ever want," said Bob. "And so we wanted somebody that we thought could handle all that—that was good under pressure." Bob had flown and worked with Sally before, and he thought she was up for it, based on her experience on the competitive tennis circuit and her demeanor in the T-38s. He also wanted to have a cohesive crew. He knew that Sally had worked extensively with John, and he'd seen them out at parties together. He was confident they'd all get along during training—and while flying in space.

There was one other key trait that Bob and George thought Sally had, which was counterintuitively appealing. As an introvert, Sally wasn't exactly one to *seek* the spotlight or fame. And both men agreed that such a personality might fit best with being The One. They didn't want to choose someone who wanted the attention too badly. "I know I felt, and I think George did also, that Sally was not the type of person [for whom] being the first was going to go to her head," Bob later explained. "She could take it in stride."

Bob pretty much agreed with George that he thought the assignment should go to Sally. But to really convince his boss, George did what any good engineer and manager would do: he made a spreadsheet. In a memo addressed to Chris Kraft, George composed a matrix of various mission specialists, including all six women and two of the Black astronauts. Next to their names, George put Xs to indicate each astronaut's skills and proficiencies.

A quick glance at the grid showed Sally and Judy deadlocked. They both had the most Xs next to their names, indicating they were great at handling various payloads and software. And in the "comments" section, George wrote that both were "extremely well qualified on RMS," or the robotic arm. Anna wasn't far behind in her X count, and a comment next to her name noted she had "outstanding public presence."

But giving Sally the edge was the appearance of one more X, indicating she had a better understanding of more systems than the other two front-runners. And with her knowledge of the robotic arm, that sealed the deal. It was a photo finish, but Sally had outpaced the other two. "It turned out Sally was the best qualified," said George. "There wasn't a male or another woman who could do it as well as she could on the remote manipulator."

George put the finishing touches on the spreadsheet and sent it up to Chris. And he waited. He was counting on the simplicity of the document to win the day. It was all there in black and white. Just a few days later, Chris finally accepted Sally and the other assignments.

The crews were set. Sally's life was about to change. It was just time to tell everyone.

Early on Monday morning, April 19, 1982, Sally was summoned to George's office. Alone, she walked toward the elevator and rode to the eighth floor, her body tingling. Sally wouldn't admit it out loud, but she had an idea of what was about to happen. Over the weekend, she'd been told to come home early from a trip to

Disneyland she'd taken with Steve, a quick trip to one of her favorite childhood places. For her, the best parts were the E tickets, which admitted them to the most advanced rides of the day.

It had to have been something important to cut short a vacation. Either she was in trouble, or it was something very good.

She walked into George's office, and George, no master of small talk, got straight to the point. "How do you like the job you have now?" he asked.

Sally was flummoxed. "Well, what is my job?" Sally replied. She'd just finished up her recent stint as CAPCOM during STS-3 and had been in limbo on what her next assignment would be. Was this just a chat about scheduling?

"We thought that maybe you enjoyed what you were doing so much that maybe you wouldn't want to fly on a crew," George said, referring to Sally's love for her CAPCOM duties.

Taken aback, Sally pondered what he meant. Was he about to offer her a seat assignment? Sally answered fast, making it crystal clear: she was *very* interested in flying on a crew. Without saying much else, George then escorted Sally out of his office and upstairs to another office. This one belonged to Chris Kraft.

The pair sat down across from the director's desk, and Chris launched into a speech about what accepting this assignment would mean for Sally. He told her it was going to be a difficult few months and years. That the job would come with unprecedented responsibilities, more than just the roles she'd perform during the actual mission. What he wanted to know, in fact what he absolutely *needed* to know, was this: Had Sally thought long and hard about what accepting this assignment would mean? The lights of the combined American media focused squarely on her? The probing questions that were bound to be asked? NASA would be there to support her, but if Sally hadn't fully considered the implications, she better think about them—fast.

Sally understood what her boss was trying to do. He was giving her an out. And this was the only chance she had to take it.

But it's against the nature of an astronaut to turn down an open seat on a spacecraft. Flying sooner is always better than later. "There was no doubt in my mind that I wanted to do that," Sally said.

Sally told Chris that she *had* thought about this. She was ready.

CHAPTER 10

Ready, Set . . .

"How does it feel to realize that because primarily of the luck of your birth, along with some good work and so on, you are going to become a footnote in history and a trivia question subject forever, among other things?"

The male reporter asked the question of Sally Ride sitting on the stage in front of him. Wearing a royal purple blouse, she sat behind a conference table next to her STS-7 crewmates. NASA had just announced Sally's flight assignment to the country a week earlier, and the agency scheduled a press conference in JSC's Teague Auditorium with the history-making crew. The press had all been jockeying with one another to talk to America's first woman to go to space, and the presser gave them an opportunity to get all their questions in at once.

"How do you relate to all of that?" the reporter continued.

Everyone in the room laughed, including Sally. Gracefully, she answered without a hint of frustration.

"Gosh, that's quite an honor," Sally replied with a laugh. "Well, of course, I was very honored that NASA chose me to be the first woman. I guess that I was maybe more excited about getting a chance to fly early than I was about getting to be the first woman."

The press conference only lasted a half hour, but nearly all the questions revolved around Sally in some way. A reporter wanted to know if Sally had always wanted to be an astronaut, and if she'd trained before she came to Houston. Another male reporter wanted to know how washing and personal hygiene would work on the Shuttle with a coed crew. "How the hell do you work around that?" he asked.

Sally graciously deferred the question to Bob Crippen. "Look, give us a chance to work at it," he said. "There's no doubt that we're going to become very familiar with one another over the next year or so." When the reporters weren't asking Sally an inane question, they were asking it of her colleagues. One wondered if the men would defer to Sally and allow her to do more tasks out of politeness, analogous to opening the door for her here on Earth. "I don't think that Dr. Ride needs anybody in any group to defer to her," John Fabian replied. "I think that her capabilities take care of themselves and she'll stand high in any group."

Sensing the overall theme of the conference, Crip made one thing clear to everyone in the room. "Sally's on this crew because she's well qualified to be here," he declared, unprompted.

But standing out among all the questions was *the* question. The one Sally would be asked for the rest of her life. The one that would ring in her ears for decades and eventually lose all meaning every time it was asked.

How does it feel?

"I'm sure you've been asked many times how it feels to be the first woman astronaut . . . How does it feel to be asked that question?" one reporter asked. "How do you respond to that question?"

"I've also been asked many times how it feels to be asked that question," Sally said, as the room laughed. "I think that I'm going to get real tired of that question."

NASA also held a press conference for the crew of STS-8, where Guy Bluford faced similar questioning from the press. The press didn't spare him *the* question either. The world wanted to know how he *felt* about being the first Black American astronaut to go to space. "It's normally the first man who does it is under a little bit of tension, so to speak, under a little more scrutiny than the people who come behind him," Guy said. "But I'm excited just to be here."

Sally and the rest of the newly assigned TFNGs were obviously exuberant about the imminent prospect of flying—well, *relatively* imminent—but those emotions weren't exactly shared by the rest of their unassigned class. George had announced the crews for the three missions to the rest of the astronauts on the afternoon of April 19, the same day Sally found out about her assignment. She and the other three members of the crew had initially been forbidden from telling anyone until later that afternoon, so they'd gone to the cafeteria together to celebrate in silence. While at lunch, Mike Mullane sat down next to them and lamented to John about not knowing when they'd all get assigned. John stayed uncomfortably quiet.

At an informal meeting of the astronauts, George put it bluntly. "We've made some crew assignments," he said, and then read out the names. Knowing that he'd just broken the hearts of most of the people in the room, he added: "Hopefully we'll get more people assigned soon."

With those words, the mood instantly mutated. A schism formed among the TFNGs, separating the class into those with assignments and those without. While the unassigned astronauts smiled wide and congratulated their lucky peers, many tried to suppress the extreme jealousy and disappointment eating away inside. Watching their classmates celebrate while their future remained in limbo was excruciating. One astronaut mumbled "This is bullshit" under his breath when they all dispersed.

For the Six, at least, the unspoken competition had officially come to an end. Sally would be The One. That was that. For Judy, Anna, and Kathy, the news came as a blow. "I would have loved to have gone first," Kathy said. "I was confident in my abilities." Sally's name would reverberate throughout history. And in addition to the obvious perks that came with being the first American woman in space, Sally would, most enviably, get to fly *before* them. When you're an astronaut, and your sole ambition is to fly, it stings knowing that you still have many months or even years to wait. That was the tough part for Shannon, who wasn't so caught up in being first. She just wanted to see space—she'd wanted it her whole life.

"Of course, every one of us wanted to be first, but then that's true of everybody in their class—for the guys, too," Anna said. "That's just the nature of the competitive group of people that have basically been competing their whole lives."

For Rhea, George Abbey's news also hurt, but she knew that her decision to have a baby was probably going to push back her flight assignments. "I had to decide: Do I want to have a baby now and be delayed in getting a flight? Or do I want to get a lot of flights and then try to have a baby? And I wasn't terribly young anymore. So, to me, it was just more important to have a baby—to have children and maybe never get a flight."

Sensing Rhea's despondency, Hoot reminded her of something

important. "I think later on in life you'll be just as happy that you got to fly, but that you weren't the first," he told her. It would be years before she understood the power of those words.

That evening, the astronauts hosted a small celebration at the Outpost for the crewmates, and the booze helped to melt away the sadness that had gripped the unassigned TFNGs throughout the day. The alcohol also manifested into some boldness. At one point after a few beers, Hoot grabbed Steve Hawley at the bar and approached George in the booth where he was sitting.

"Mr. Abbey, Stevie and I need to tell you something," Hoot began, his arm around Steve's shoulder.

"What's that?" asked George.

"You really screwed up today."

"How's that?"

"Because you didn't name us!" said Hoot. Completely blind-sided by Hoot's proclamation, Steve gave a look that read *I'm not with him*, and tried to squirm away from underneath Hoot's arm. George simply chuckled.

The next day at work, Rhea and Hoot ran into another TFNG who was absent from the festivities the night before. "We didn't see you at the celebration last night," Hoot said to their colleague.

"Well, I didn't see anything to be celebrating," the astronaut said glumly, before sauntering off.

Carolyn Huntoon had expected Sally's assignment would be tough for the other TFNGs to swallow, as well as for her five female coworkers. She called up Sally soon after the announcement.

"You got it," said Carolyn over the phone.

"Yeah," replied Sally.

"Now, be nice."

"Yes ma'am."

Sally did her best to be outwardly humble, while inwardly, she did backflips. But her excitement mostly stemmed from the fact that she was going to get to *fly in space*. When she talked to Steve about the assignment, all her joy revolved around the fact that she had crewmates—all friends of hers—and that she got to start training as soon as possible. But Steve noticed something else. She didn't seem to acknowledge the implications of being first. He knew she didn't like being a public figure, and he wondered if she fully understood the amount of pressure that would be placed on her in the years to come. "She saw all of that [having crew members, getting to fly] as being really exciting, and I thought that was wonderful," said Steve. "But I wondered, is that going to come in conflict with the burdens that go with the assignment? And how is she going to deal with that?" He didn't say anything to her at the time, but he figured those burdens would creep up soon.

There wasn't much time to dwell on it anyway. Sally's training began immediately, with the entire crew of STS-7 moving into John Fabian's former office. (John's *Playboy* pinup calendar had mysteriously disappeared before Sally arrived.) The team got to work, mapping out the flight plan and the payloads they'd be working with. They had two satellites to deploy; and flying around the SPAS-01 satellite would be the mission's showstopper. Sally's robotic arm expertise had been key in securing her a spot on the crew, and she had to keep her skills sharp. That meant continuing her trips to and from Canada to train with John, as well as manipulating the controls in training sessions and simulators in Houston.

Sally would also be performing another crucial role, courtesy of Crip. He'd made her the flight engineer, a critical rank that made her the backup to Bob and Rick. Sally would sit just behind the commander and the pilot during the ascent to space and the return trip home—the two most pivotal legs of the mission—and if something went wrong during those portions of the flight, Sally would have to step in. She'd help with the checklist or monitor the displays, so that Bob and Rick could focus on saving the crew. The role meant that Sally spent plenty of time in the simulators with Bob and Rick, practicing the ascent and reentry over and over and over again. The role also meant that Sally would be afforded one hell of a view during the climb to space: a shot of the windows and control panel between the pilot and the commander.

She'd be able to catch a glimpse as the thick blue atmosphere of Earth petered away, transforming into the inky black void.

———

IN JULY 1982, just a few months into training, Sally took her mind off work for a few days to focus on another important task. One weekend, she and Steve traveled to Kansas to Steve's parents' house, a trip they'd made many times before. But that Saturday, Sally's sister, Karen, nicknamed Bear, was there. Steve's father and Bear, who'd since become a Presbyterian minister, co-officiated Sally and Steve's wedding out in the backyard, beneath a clear midwestern sky.

It was a small affair, something that came as a bit of a surprise to everyone—even Steve. Leading up to the wedding, he and Sally often didn't see much of each other as a couple. With their hectic work schedules, they could resemble two ships passing in the night. Sally would often fly to Canada for weeks at a time; meanwhile, Steve was frequently traveling to and from Cape Canaveral to support various Shuttle missions. And when they *did* find themselves in the same place, Steve would want to spend his time with Sally while Sally would opt to go flying for the weekend with one of the TFNG pilots.

"It didn't seem like I was a real priority," Steve would later say, looking back. "And, in fact, I was going to suggest: why don't we go separately from here on out."

One night, he prepared to bring up the idea but before he could say anything, Sally shocked him with a declaration. She told him that her feelings for him were actually quite strong. *Okay, well I missed that*, Steve thought. He knew he still loved her, even if he felt his feelings for her eclipsed the supposedly strong feelings she had for him. So they continued dating until one day, Sally casually suggested the idea of getting married.

Steve indicated it was mostly up to her. "I knew back then, and it became even clearer later, that she was pretty much going to do what she wanted to do. And in terms of our relationship, I guess I decided that was okay with me. However she wanted to play it was fine."

One thing led to another until the two were exchanging vows in that Kansas backyard, the pair wearing T-shirts and white jeans.

Both publicity shy people, Sally and Steve knew they didn't want an elaborate event. "We didn't want to make a big deal of it," Sally said. Steve gave permission to Dick Scobee to let the astronauts know at the Monday meeting after the wedding. And ultimately, playing it low-key achieved the desired result. Their wedding only manifested as a short paragraph in the newspapers.

Soon the press had other NASA relationship news to focus on. Just as Sally and Steve turned a new chapter in their lives, so did Rhea and Hoot. The couple had just bought a new house and were set to move there in late July, just a week before Rhea was due to give birth. But as Rhea packed up boxes with dishes and knickknacks one Saturday morning, rolling cramps began to grip her. She knew she was going into labor—a week early.

Rhea labored for hours at Clear Lake Hospital, and after an exhausting day that lasted until four in the morning, the doctors finally opted to do a C-section, noticing signs of distress in the baby. In the wee hours of July 26, 1982, Paul Seddon Gibson came into the world. But as soon as he took his first breaths, both Rhea and Hoot could tell something was wrong. He wasn't crying and his skin was turning blue. The doctors realized that the newborn needed swift emergency care. Paul had swallowed meconium, stool that he'd ejected into the amniotic fluid during the intense labor. With his lungs coated in the waxy fecal substance, he needed help to breathe—fast. The doctors put him on 100 percent oxygen inside a protective bubble as they rushed him via helicopter to the NICU at downtown Hermann Hospital.

The next few days would be filled with painful anxiety as the couple prayed Paul would come through. At one point, Rhea's dad, who'd come to be with the family in Houston, asked a doctor if survival would be an issue, with Hoot standing alongside. The doctor solemnly replied, "Yes," to Hoot's horror. But after days of intense therapy and treatment, little Paul started breathing on his own and Rhea, who also needed to heal from her labor, traveled up to Hermann Hospital to hold him for the first time.

With their now healthy baby in their arms, Rhea and Hoot presented Paul to the press during a conference arranged by NASA. Everyone wanted to get a glimpse of the world's first "astrotot," the first child of two astronauts. But he wouldn't be the only one for long.

At that point, Anna was thirty-three and Bill was thirty-six and the couple realized that the window to have children without risking unwanted complications was closing. Also, their flight assignments were taking much longer to materialize than they had anticipated. When they'd first started the program, they were told they'd be flying within three years. Now, nearly five years later, Anna still didn't know where she stood in the flight lineup.

"I wasn't assigned to a crew, so for me it was a scary decision," she said of getting pregnant. "I didn't have any real confidence I was helping my career. But I felt it was important enough."

Just months after Paul's birth, Anna found out that she and Bill were expecting. And as Rhea had done, she decided to keep it a secret from NASA as long as she could. She'd seen what had happened to her colleague, and she didn't want to be grounded from the T-38s.

Keeping silent turned out to be harder than expected, though. For one Shuttle mission, Anna was at Kennedy Space Center and assigned to be Astronaut Support Personnel, tasked with strapping in the crew before they launched. While that job was easy enough, the assignment also meant she had to prep for emergency scenarios. If something happened to the crew during the early parts of the launch, she had to be the one to haul them out of the cockpit. NASA wanted Anna to test her ability to do that leading up to the flight. At her diminutive height, it was always going to be a struggle carrying two-hundred-pound, six-foot-tall men out into the Florida heat. But at the time, Anna was secretly four months pregnant, too.

"Anna, if you do that, I'm going to shoot you," a concerned Bill told his wife when he found out.

"Bill, if you can think of a way for me to get out of it without telling them I'm pregnant, fine," Anna told him. "Otherwise, I'm going to do it."

As promised, she did her job as asked. During a test at the space center, Anna lugged two male crew members, one from the pilot's seat and the other from the commander's, working alongside a female suit technician. It had been tough, but she got through it just fine with no complications. However, she wasn't quite able to keep her secret. After the body dragging, the tech wondered to her team if they all noticed Anna "had a baby bump going on."

It wouldn't be long until Anna had to come clean. Her belly was just getting too big to explain away. When she finally did spill the beans, the person who'd assigned her to the body-dragging task wasn't happy.

ON AUGUST 19, 1982, a woman with straight red hair and clad in a Soviet space suit climbed into a Soyuz capsule more than seven thousand miles from Houston in Kazakhstan, before blasting into orbit. Her name was Svetlana Savitskaya, and with her liftoff, she became the second woman in the world to fly to space. When the Soviets learned of NASA's recruitment of nonmilitary astronauts, the nation began another selection round for female cosmonauts. And when Sally's assignment was announced, the urgency to fly another woman cosmonaut grew. Svetlana made the cut, flying nineteen years after her predecessor, Valentina, and docking with a Soviet space station called Salyut 7. When she arrived at the station, the two cosmonauts already living there presented her with flowers that had been grown in orbit. When the three had their first meal together, they presented her with an apron.

"There is a kitchen and that will be where you work," cosmonaut Valentin Lebedev jokingly told her.

Svetlana's flight meant that Sally would become the third woman to fly to space, but it was a demotion that barely registered for her. As 1982 transitioned into 1983, her training gradually became busier and more complex. A new addition to the team late in the game had also changed the crew dynamic.

George had assigned another TFNG, Norm Thagard, to the mission toward the end of 1982, rounding STS-7 out to a crew of five. A medical doctor by training, Norm was tasked with investigating something that had been plaguing the astronaut corps since human spaceflight began: a condition dubbed Space Adaptation Syndrome (SAS). More than half of the people NASA sent into space would get unbelievably nauseous and sick, vomiting profusely as soon as they got into orbit. And the condition seemed to strike at random. Some of the most seasoned pilots would feel their stomach churning the moment they were in microgravity while the mission specialists who normally shied from thrills on the ground leisurely floated, vomit-free. Norm's task was to develop experiments to determine who got sick, why, and how to stop people from getting sick.

After he joined the crew, the STS-7 astronauts began spending more and more time in the Shuttle simulators, alternating between the ones that stayed fixed and the motion-based simulator that shuddered and convulsed, mimicking the turbulence an astronaut would feel in the real Space Shuttle. Some of the astronauts practiced flipping just the right switches that would deploy the satellites while Sally and John traded off working with the robotic arm. Every punch of the joystick, every push of a button had to be expertly choreographed on the ground so that it became second nature in space.

The simulations were short at first, but over time, they grew to last hours. Eventually, the team graduated to integrated simulations, grueling days-long rehearsals that involved not just the crew but whole swaths of engineers and flight controllers in Mission Control. These sims, which could last up to fifty-six hours, meticulously re-created the biggest moments of the flight. To keep the crew on their toes, the Sim Sups were always there, ready to mix up a cocktail of glitches and malfunctions that could tear apart the mission. Each time, Sally and her team raced to fix the problem before their virtual Shuttle fell to its doom.

The days became long and daunting, but the crew and training teams found ways to make them fun. Once, before Sally and the crew were scheduled to strap into the motion simulator, a few pranksters on the training team got a rubber mouse and tied it to some fishing cord, affixing the string and fake rodent to the cockpit's glare shield where they couldn't be seen. When the STS-7 crew was inside, the simulator started to rotate ninety degrees, rolling the crew onto their backs to simulate their positions when they climbed to space. As gravity took hold of the mouse, it tumbled out from its hiding place and landed right in front of Sally's face, as had been intended. The cord had been cut at just the right length so that the mouse would dangle inches from her nose. Sally let out a loud scream and the training guys snickered in delight. She eventually took the mouse and refused to give it back.

Another day, Sally had to do her standard bench checks, which meant reviewing the personal items she wanted to take to space. She'd opted to fly with some personal effects, which needed to be approved. But the engineers wanted her to look over some other items that had been laid out for her.

On her bench was a personal hygiene kit that NASA engineers had created. Such kits harkened back to Apollo and Gemini. They typically included things like toothbrushes and shaving cream, your basic grooming supplies. But now NASA had made one geared toward women. It included standard things to keep yourself fresh each day but nestled inside was a smaller kit wrapped in yellow plastic.

It was a makeup kit.

NASA personnel had actually been working on the kit for a while. One can't simply launch a tube of lipstick into orbit without planning. There were procedures for this type of thing. Each product had to go through a small amount of testing, to see if it was flammable or if any chemicals leached off the object while in space.

One of Anna's first technical assignments had been to help decide what should go in the kit for women. She didn't know much about the best skin-care or hygiene regimens; once she started working crazy hours in medical school she'd resolved not to wear any makeup. She didn't like the idea of falling asleep at some odd hour and waking up with mascara streaming from her eyes. But one thing she did remember was Nivea cream. Her German mother had used it frequently, so she managed to get that inside the kit.

To Sally, the idea of having to even think about wearing makeup in space felt like a joke. "A makeup kit brought to you by NASA engineers," she said, adding, "You can just imagine the discussions among the predominantly male engineers about what should go in a makeup kit." She knew she needed reinforcements on this, so she looked around for an ally. Luckily, she ran into Kathy and with pleading eyes asked her to take a look.

"I think she would have grabbed whichever one of the several of us she saw at the time," Kathy said. "It was a moment of convenience thing."

Together, the two did their best to approve the hygiene and makeup kits, even though it was the last thing they wanted to do. The pair may have rolled their eyes, but the idea of wearing makeup in space wasn't universally derided by the Six. NASA had asked the women when they first arrived if makeup was something they might want. Most shook their heads, but Rhea actually spoke up. "If there would be pictures taken of me from space, I didn't want to fade into the background, so I requested some basic items," she said later.

While reviewing the products, Sally noticed a weird band of pink plastic. She put her hand inside the hygiene kit and tugged on it gently. Out came a tampon. And then another. And then another. Strung together like sausages, the tampons formed a never-ending caravan. Both Sally and Kathy stifled their laughter. The engineers present at the bench check asked Sally if one hundred would be the right number of tampons for her weeklong trip.

"No. That would not be the right number," Sally replied calmly.

"Well, we want to be safe."

"Well, you can cut that in half with no problem at all," said Sally while Kathy devolved into giggles.

WHEN SALLY WAS first introduced to the country, NASA had concocted a plan to guard her from overwhelming press attention during her training. The space agency often used her busy schedule as an excuse to turn down media opportunities, and when Sally did appear in public or in front of reporters, she had backup. Crip worked with NASA's public affairs office to ensure that Sally always had at least one member of the crew with her—and preferably the whole crew—when she was speaking on behalf of the launch. As the flight grew closer, the press attention grew more oppressive.

The press, however, were getting antsy for Sally's story and beginning to show their frustration with NASA. So to satisfy the public's thirst for knowledge, she agreed to cooperate with a reporter from the *Washington Post* for a detailed series of articles. This particular journalist was someone Sally felt comfortable sharing her life with—because she'd already been doing it for years. The reporter was Sue Okie, her childhood friend, who'd written for the *Post* for a few years before pursuing medicine.

To Sally, Sue didn't *feel* like a journalist. She was just Sue, and the astronaut felt at ease talking to her longtime companion for hours on end. However, there'd been a moment a few years prior when Sue's profession had reminded Sally of the price of being a public figure.

While working for the paper, Sue had made it known to the sports reporter there that she and Sally were close. The reporter, who knew a lot of tennis players, shocked Sue by saying that she knew Sally and her college roommate, Molly, had been a "lesbian couple." The declaration was news to Sue. As both a journalist and a friend, Sue wanted to get confirmation from Sally herself. One day, just a few years after Sally had been selected by NASA, Sue traveled up to New York to see her astronaut friend, who was giving a speech. After the event, the two were in Sally's hotel room together when

Sue built up the courage to ask. She knew that Sally and Molly had been close, but had they been in a romantic relationship?

Sally paused before she answered. "No," she said. "Molly wanted that, but I didn't." That put the matter to rest for Sue, who never asked Sally about Molly again. But for Sally, the fact that Sue had known about the relationship sent a shock through her system. *Who had told her? Was it another journalist? Did Molly say something?* As someone who didn't like to share even the most mundane parts of her life, Sally grew angry at the idea that the world might find out about this very personal part of her past.

At the time, attitudes were changing about gay couples but not fast enough. Sally had watched her idol Billie Jean King come out about her relationship with another woman in 1981, and the repercussions had been swift. While King received support in the press, within twenty-four hours of going public the tennis star lost all her endorsements.

Sally could only imagine what would happen if her relationship with Molly surfaced in the news before her flight. The idea incensed her, and she tried to figure out how to prevent that from happening. But time passed, and no rumors surfaced. Eventually Sally got her assignment, married Steve, and became too distracted to worry about events that belonged to her pre-NASA past.

Despite that personal question in a New York hotel, Sally trusted Sue not to reveal anything too private. So in the months just before her flight, Sally met with Sue to go over her life and training. Together, they toured the National Air and Space Museum in Washington, D.C., taking in the relics of spaceflight history. While gazing at the old Apollo capsules and vintage space suits, Sue reminded Sally that her name and visage would appear in this very building one day. "They'll have your picture in here along with Amelia Earhart's." Sally grew beet red and turned away. "That's ridiculous!" she cried.

"She couldn't talk about that or even think about that," Sue said. "It was just too weird for her."

The one thing Sally did like talking about was her training. Sue hopped a flight down to Houston to witness Sally's expertise in action. She watched in awe as her friend, the former self-proclaimed "slacker," pored over her mission manuals and flight books deep into the night, studiously preparing for the launch ahead. "I have lost my dominant trait, which has been not to work at things," Sally told Sue. "I'm really working hard, and I have been for three years. And I enjoy it. In fact, I'm obsessed with it." At JSC, Sue watched her friend join her crew to combat every manner of failure possible that Mission Control threw at them in the simulator. During one training run, the crew did every procedure they possibly could to overcome the swarm of virtual failures, and the Space Shuttle still crashed, theoretically killing them all.

That moment shocked Sue. It served as a wake-up call that spaceflight wasn't a risk-free enterprise. The last major spaceflight failure Sue remembered had been Apollo 13, and the crew had survived that one. After the Shuttle conducted its fourth flight, NASA had declared the vehicle "operational," implying that all its subsequent flights would be routine and safe. That her friend might die in some fiery explosion sent shivers down Sue's spine. But Sally, as was her way, didn't betray any fear at all. She and the others would shrug off a crash and vow to do better on the next run. "They can always give you enough failures that the plane will crash," Crip told a stunned Sue afterward.

Sue's stories ran for four days in the *Washington Post*, but that didn't sate the country's curiosity about America's first woman in space. At one point, Sally and her crewmates made their way to Washington to have lunch with President Reagan—not someone Sally was then a big supporter of. In a perfect twist of fate, both Bob and Sally became stuck in the elevator on their way from NASA Headquarters to the meeting. Bob picked up the elevator phone and hollered to whoever was listening, impressing on them that they had an important meeting to attend at the White House. Thirty minutes later, the duo was drinking tea with the president in the Oval Office.

"It seems my life isn't my own anymore," Sally admitted, and it may have been true. Sally's smiling face appeared on the covers of *Newsweek, U.S. News & World Report, People*, and more, along with stories that showed journalists doing their best to pry information out of Sally and her family. Her sister, Bear, admitted to *Newsweek*: "She doesn't offer information . . . If you want to know something about Sally, you have to ask her." However, the writer lamented that Bear had never tried interviewing Sally herself. During an appearance on *Today*, host Jane Pauley asked Sally if she thought that, because of her gender, there'd be more scrutiny on her relative to her crewmates. "It seems to me I ought to be asking *you* that question," Sally responded. A *Los Angeles Times* article said that reporters found her "difficult, unreachable, stone-cold, in contempt of the press." Sally just refused to play all their games.

Legendary anchor Tom Brokaw dedicated an entire segment to Sally on NBC News, hoping to paint a more intimate picture of America's first woman in space. With a camera crew in tow, he met with the Ride family in California in an attempt to glean more about

"You know, those guys with the big hats," the reporter said.

"Well, I'm a native Texan," said Bob. "I wear a big hat, and I drive a pickup truck, and I think I disagree with those guys in the big hats. I think it's great that there are women astronauts, and I think it's great that Sally Ride is making this flight."

Rick Hauck shot down the idea that he was somehow jealous of the attention Sally received.

"I didn't join this program to get media attention, but now that I'm flying with Sally I'm getting too much," he said.

The press jabbed and the crew parried, back and forth throughout the session. They seemed to get into a good rhythm, but nothing could have prepared Sally—or any of the crew—for a question asked of Sally by a reporter from *Time*.

"During your training exercises as a member of this group, when there was a problem—when there was a funny glitch or whatever—how did you respond? How did you take it as a human being? Do you, do you weep? What do you do?"

Sally's face screamed what she couldn't say out loud herself: *You cannot be serious.* Trying her best to remain as collected as possible, she laughed, shook her head, and gave the best response she could think of: "Why doesn't anybody ask Rick those questions?"

The crowd laughed, while Bob jumped in again to help. "The commander weeps!" he joked.

Laughing, Sally gave a more serious reply to the reporter. "I don't think I react any differently than anybody else on the crew does."

Sally did get a short moment to sum up how she really felt about all this attention. "It's too bad that our society isn't further along and that this is such a big deal," she said. "But I guess if the American public thinks that it's a big deal, then it's probably good that it's getting the coverage that it's getting. I think it's time we get away from that, and it's time people realized that women in this country can do any job they want to do."

If Johnny Carson realized that, there was no evidence in his nightly *Tonight Show* monologues. He continued to mine humor— or what passed for it—from Sally's gender. One night, he joked that Sally "just canceled her Space Shuttle flight because she didn't want to be seen in the same outfit for six days." He blamed the two-month delay of the mission on Sally being unable to "find a matching purse to go with the space shoes." Many in the audience met his jokes with groans.

America's opinions had swelled into a cacophony that seemed almost impossible to escape. But finally, a week before her flight, Sally and the rest of her crew entered quarantine, staying in a trailer that was itself located inside a JSC building near the gym. She couldn't have listened to the public's opinions if she'd tried. And at L-3, or three days before her scheduled liftoff on June 18, she climbed into the back of a T-38 and jetted east to meet the space shuttle *Challenger*, waiting for her in Cape Canaveral.

The first six women astronauts selected by NASA gather around the "personal rescue sphere." During the selection process, all the women had to curl up in the sphere to prove they weren't prone to claustrophobia. NASA/INTERIM ARCHIVES/ GETTY IMAGES

All six of the women astronaut candidates, as well as their male peers who weren't jet pilots, had to undergo water survival training in Florida. Anna Fisher and Sally Ride can be seen sitting on the dock, waiting their turn to be pulled into the air by a helicopter. KEN HAWKINS /ALAMY STOCK PHOTO

From left: Sally Ride, Judy Resnik, Anna Fisher, Kathy Sullivan, and Rhea Seddon pose at Homestead Air Force Base in Florida during water survival training. SPACE FRONTIERS/ ARCHIVE PHOTOS/ GETTY IMAGES

LEFT: This photo, taken of Anna Fisher during a space-suit fitting, became iconic. JOHN BRYSON/SYGMA VIA GETTY IMAGES

BELOW: Kathy Sullivan stands before the nose cone of a WB-57F reconnaissance aircraft at Ellington Field Joint Reserve Base on July 1, 1979. She'd go on to set an unofficial altitude record for women while flying in the aircraft, reaching an altitude of 63,300 feet. SPACE FRONTIERS/ARCHIVE PHOTOS/HULTON ARCHIVE/GETTY IMAGES

LEFT: Sally Ride climbs into a T-38 jet. Astronaut candidates were required to get fifteen hours a month of flight time in the jets. Mission specialists like Sally flew in the back seat with pilots in the front, but each of the Six learned how to fly and land the fighters themselves. NASA/AFP VIA GETTY IMAGES

In the foreground, George Abbey (center) stands next to STS-1 crewmember Bob Crippen (left) and astronaut Joe Engle (right) at NASA's Kennedy Space Center in Florida. The photo was taken on April 11, 1981. NASA

Sally Ride and Anna Fisher work together at Kennedy Space Center on the payloads for Sally's flight STS-7. Anna, who was the lead Cape Crusader for STS-7, is visibly pregnant with her first daughter, Kristin. NASA PHOTO SCANNED BY J. L. PICKERING

The new Space Shuttle *Discovery* is transported to its launch site in Cape Canaveral, Florida, on the back of a modified Boeing 747. *Discovery*'s first flight to space would be STS-41-D, Judy Resnik's first flight. NASA PHOTO SCANNED BY J. L. PICKERING

LEFT: The crew of STS-41-D poses for a photo mid-flight. Pictured clockwise from Judy Resnik are Steve Hawley, Mike Coats, Henry "Hank" Hartsfield, Mike Mullane, and Charlie Walker. NASA PHOTO SCANNED BY J. L. PICKERING

ABOVE: The unfurled solar array carried on STS-41-D. Judy Resnik was in charge of the experiment while on board. NASA PHOTO SCANNED BY J. L. PICKERING

LEFT: Kathy Sullivan in her flight suit, undergoing preflight checks. NASA PHOTO SCANNED BY J. L. PICKERING

Kathy Sullivan and Dave Leestma perform the orbital refueling system experiment during a spacewalk on STS-41-G.
NASA PHOTO SCANNED BY J. L. PICKERING

Astronaut Dale Gardner holds up a "For Sale" sign during STS-51-A, after he and Joe Allen successfully retrieved two stranded satellites in space and stowed them in the Space Shuttle's cargo bay. Anna Fisher aided the recovery while working the robotic arm. NASA PHOTO SCANNED BY J. L. PICKERING

Anna Fisher is reunited with her husband, Bill, and daughter Kristin after returning from space during STS-51-A. NASA PHOTO SCANNED BY J. L. PICKERING

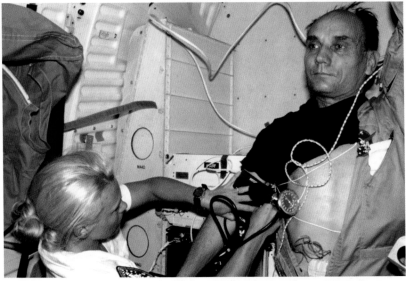

Rhea Seddon performs a medical experiment on Senator Jake Garn during STS-51-D.
NASA PHOTO SCANNED BY J. L. PICKERING

The crew of STS-51-D pose with their makeshift creations for recovering the Syncom satellite, the "lacrosse stick" and "flyswatter." Clockwise from the top left is Jake Garn, Donald Williams, Karol "Bo" Bobko, Charlie Walker, Dave Walker, Rhea Seddon, and Jeff Hoffman.
PHOTO SCANNED BY J. L. PICKERING

Rhea Seddon is reunited with her three-year-old son, Paul, after returning to Houston from her flight STS-51-D.
AP PHOTO/ED KOLENOVSKY

LEFT: Shannon Lucid floats near the cockpit during her flight STS-51-G. NASA PHOTO SCANNED BY J. L. PICKERING

ABOVE: Before their flight, Judy Resnik poses with her female crewmate, Christa McAuliffe, who was set to become the first teacher to fly to space. NASA PHOTO SCANNED BY J. L. PICKERING

LEFT: Prior to the launch of STS-51-L, the extremely cold weather caused icicles to form on the launchpad structure, leading to concerns that the ice could break and damage the Space Shuttle during launch. NASA PHOTO SCANNED BY J. L. PICKERING

At 73 seconds after launch, a "major malfunction" occurred during STS-51-L, marking the final flight of the space shuttle *Challenger*. The accident led to the deaths of all seven crew members on board, including Judy Resnik, one of the Six. NASA PHOTO SCANNED BY J. L. PICKERING

The remains of the *Challenger* crew members are transferred from seven hearses to a transport plane at Kennedy Space Center. NASA PHOTO SCANNED BY J. L. PICKERING

Sally Ride's life partner, Tam O'Shaughnessy, accepts the 2013 Presidential Medal of Freedom on behalf of Sally, who died of pancreatic cancer on July 23, 2012.
LEIGH VOGEL/WIREIMAGE

CHAPTER 11

Sally's Ride

eorge noticed that Sally was pacing.

They were standing together in Cape Canaveral, just days before Sally's flight was set to launch. She and the rest of the STS-7 crew were making use of NASA's designated beach house, a dilapidated, boxy condo nestled in the sand just a few miles from Kennedy Space Center's two main launchpads. When NASA purchased the land for the space center, the agency razed most of the site's beachfront property, save for one lone condo. As the Shuttle began flying, the sole beach house evolved into a small sanctuary for the crew scheduled to fly next—a place to unwind in the days before liftoff.

Not one to openly reveal her emotions, Sally didn't express any outward anxiety to anyone. But the pacing worried George. He knew that everyone must be a little nervous just days before getting ready to ride on a controlled bomb to space. It'd be strange if they *weren't*. "You're going to feel uneasy certainly the first time you do it," said George. But he also wanted his crews to be as calm and relaxed as possible before flight—so he recruited some help.

Sally had invited nearly everyone she knew to the launch. The Ride family had crossed from one coast to another, as had plenty of Sally's old tennis buddies—including Tam O'Shaughnessy, whom she'd stayed in touch with on and off since their days dancing to records after matches. The guest list also included some previous partners. Bill Colson got an invite, as did Molly Tyson, though the world only knew Molly as Sally's former roommate and close friend.

That's how George thought of Molly when he sought her out and asked if she'd talk to Sally for a minute. "If anyone could make her feel more comfortable, we wanted to make Molly certainly available to her," he'd say later. It wasn't a simple request. Sally and the other crew members were technically in the standard weeklong quarantine preceding flight. The idea was to limit the chance of crew members catching a nasty bug and bringing it with them to orbit.

But George decided to fudge the rules a bit. He had Molly undergo a quick physical so she could see Sally at the beach house. It wasn't a major procedural compromise. Truthfully, the quarantines were mostly in place so that the astronauts didn't interact with any children, who were walking human petri dishes.

When Molly arrived at the beach house, Sally got the unique privilege of actually seeing a member of her entourage. That was the biggest irony of inviting someone to see you fly on the Space Shuttle. Sally didn't actually get to interact with any of the people who'd traveled across the country to watch her fly. But she was pleasantly surprised to see her old friend. They talked briefly about inconsequential things, and Molly noticed that her friend seemed particularly upbeat, ready to charge into the ring. It was a nice, diverting visit for the two.

But as Molly got up to leave, Sally showed a rare moment of vulnerability.

"I'm aware that this is not without risks," Sally told her friend, her demeanor suddenly changed. "I realize I could die."

Shocked by the statement, Molly realized in that moment that the legendary Sally Ride—the woman she'd always perceived as some kind of godlike figure—could get scared, too.

Molly left, and the two never spoke about the conversation again.

———

AT 7 A.M., Sally Ride lay horizontally in a metal seat inside the space shuttle *Challenger*'s cockpit, strapped in after a whirlwind of morning prep. The day had started early, with a wake-up call at 3:13 a.m. in the crew quarters. Sally donned a striped polo and pants in a surreal daze and tried her best to act like a normal person as she trotted out to eat the traditional prelaunch breakfast with the rest of her crewmates, video cameras capturing every bite she took. "I was struggling very hard . . . trying to look like nothing unusual was about to happen to us," Sally recalled. Her meal consumed, she slipped into her sky-blue long-sleeved bodysuit that she'd wear for flight.

When the crew drove out to the pad that morning, they dropped off George at the Launch Control Center. At that point, Commander Bob Crippen turned to his crew. "That's the last one of those," he said, referring to all the officials they'd just left behind. "All the sane people are back there. We're the only people crazy enough to go out here."

Alone with their thoughts, the crew ascended in an elevator at the launchpad's base that would take them to the cockpit. A small toilet had been placed at the tower's top in case anyone had to take a last-minute nervous pee.

Sally might have needed an extra nervous pee if she knew just how many people were about to watch her take flight. On the beaches a few miles from the launchpad, the massive crowds of more than 500,000 people were up and awake, vibrating with anticipation. Among the throngs stood various celebrities who'd trekked out to Cape Canaveral to watch history being written. Gloria Steinem, the famous American feminist and activist, was standing in the nearby crowds with actress and activist Jane Fonda. Later, Jane's presence at the launch would cause a stir at NASA, after she was quoted heavily in the press about the event. Headquarters would receive an angry call from the White House, which wasn't a fan of Jane's politics, and a NASA press officer would resign over the "controversy."

But Sally was blissfully unaware of the hordes of gawkers jockeying for the best view on the nearby shores. Instead, in those last few hours the Shuttle's cockpit had become her entire world as NASA's countdown clock ticked down to zero. She stared ahead, focused on the control panel and screens that filled her view between the pilot and commander seats. About thirty minutes before launch, she looked over at John Fabian lying next to her and noticed that he was checking his pulse. Inspired, she put her fingers to her wrist to do the same. Surprisingly, her heartbeat was steady, between fifty-two and fifty-six beats per minute. She may have been nervous, but it didn't seem to translate to her cardiovascular system. She continued checking her wrist periodically as T-minus 0 came closer.

T-minus 30 seconds . . .

NASA's flight controllers handed over control of *Challenger* to the vehicle's computers. The spaceplane was now in charge. If anything went wrong, it was on the Shuttle's computers to step in and stop the flight. At that point, almost in slow motion, the seconds fell away.

T-minus 10 seconds . . .

Sally could feel the slightest bump in her heart rate, but probably only between sixty to sixty-five beats per minute. She stayed cool, focused on the moment that would completely change her life.

T-minus 6 seconds . . .

Challenger's three main engines ignited at the vehicle's base, breaking the silence and sending intense vibrations up through the cockpit. A deafening roar accompanied the ignition, the engines notifying the crew that it was almost time to break free of this earthly launchpad. "We have ignition!" Crip called out to his crew. As the astronauts had been told to expect, the entire Space Shuttle rocked backward, the so-called *twang*, making them feel as if the entire stack could topple over. But almost as quickly as it leaned, the Shuttle snapped back into place. And then it careened into the sky.

"We have liftoff!" Crip called out.

In that moment, Sally knew she was no longer in the driver's seat of her life. "All of a sudden, I felt totally helpless, totally overwhelmed by what was happening there," she said later. "It was just very, very clear that for the next several seconds, we had absolutely no control over our fates."

Completely at the mercy of NASA's army of engineers and mechanical prowess, Sally surrendered herself to the space shuttle *Challenger* as it burst forth through the clouds, thrusting its precious cargo into the stratosphere. The Shuttle twisted through the air, rolling onto its back as it took a glorious swan dive into the sky. The turbulence of launch reverberated throughout the cockpit, but Sally couldn't let herself get distracted by the intensity of her surroundings. She still had a job to do. Her main priority during flight was to keep an eye on the checklist attached to her knee, reading off the key milestones of the flight as each occurred. It took all of Sally's strength to make her first callout, squeaking out "LV, LH" when prompted by her commander just a few seconds into launch, an acronym that indicated the Shuttle was in the correct orientation it needed to be in the sky. "I'll guarantee that those were the hardest words I ever had to get out of my mouth," Sally said later.

Shuddering violently, *Challenger* pierced through the atmosphere, jostling the crew as they held on to their metal seats. It was a completely overpowering experience, but Sally felt that the months and months of training had prepared her for the sensations. What she didn't quite expect, however, was the bright flash at about two minutes after liftoff. A cluster of motors on the solid rocket boosters had fired, triggering the giant candles to break apart from the Shuttle as planned, their jobs complete. The burst of gleaming light through the windows was much more intense than Sally had prepared for, giving her a brief jolt of adrenaline.

In that instant, the violent convulsions of launching disappeared and the ride suddenly became smooth, as if the crew were on an electric train gliding on invisible tracks to space. *Challenger*'s engines propelled them the rest of the way into orbit, before they abruptly shut down a few minutes later.

"We have MECO," Mission Control announced, referring to main engine cutoff.

With the engines suddenly quiet, they were floating. Sally took the checklist attached to her and stuck it out in front of her. It did not "fall," but instead drifted leisurely through the air. Next to her, John let out several loud whoops in celebration.

This was it. They were in space.

Sally stayed strapped to her seat, waiting for the last few milestones. A loud clang sounded inside the cockpit, heralding the departure of the massive external tank as it broke apart from the Shuttle's belly and fell back toward Earth. Then it was time for the Shuttle's thrusters to do the last few burns that put them into their intended orbit. The second burn shoved Sally back into her seat. But once those burns were done, she pushed herself out of her metal chair and eventually floated toward the central control panel, getting her first look out the windows at the miraculous view. From nearly two hundred miles up, the crew watched as the continent of Africa passed below them. It had taken them just eight and a half minutes to cross the entire Atlantic Ocean.

Mission Control checked in with Bob to make sure they were all in good spirits after reaching orbit. After he replied, Sally got hold of the radio for just a few seconds to describe her experience.

"Ever been to Disneyland?" Sally asked Houston over the intercom.

"Affirmative," the mission's CAPCOM replied.

"That was definitely an E ticket."

"Roger that, Sally."

―――――――

BEFORE SHE'D TAKEN flight, Sally had told her friend Lynn Sherr in an ABC interview that her biggest concern was that she was going to somehow mess up. Unspoken in that confession was something that the Six had all been grappling with: just one mistake, no matter how small, came with outsize consequences. If Sally couldn't get it right in space, others would think that surely *all* women would falter in orbit, too.

Fortunately for future generations of women spacefarers, Sally performed just as NASA and George had hoped—though it took some time at first to adjust to working in microgravity. The first day in orbit, she felt clumsy and inefficient, needing extra time to stow equipment in her locker without accidentally pushing her body off in the wrong direction. The first couple of meals were also a pain since she couldn't anchor herself down properly and fully relax while eating with her spoon and scissors. Later, she'd talk about the "real steep learning curve towards figuring out how to operate in Zero-G." She also found herself fighting off a strong urge to simply close her eyes and drift to sleep, feeling an inexplicable exhaustion that sometimes comes with spaceflight.

But after those first few days, she adapted and took to microgravity with ease. She didn't feel one ounce of nausea, though she did get a "pitching" sensation whenever she looked out the window and saw Earth streaming by beneath her. Some of her crew members did experience bouts of nausea. Since Norm's big assignment was to learn more about Space Adaptation Syndrome, he roped Rick Hauck into some stomach-lurching experiments, tasking him with watching a spinning cylinder with black stripes on it. "That really messed my brain up," Rick said. "And at one point I said, 'Hey, I'm going to go off into the airlock and close my eyes and try to feel normal.'" Though Sally's stomach remained calm, she still opted to stay "upright" with her head pointing toward the Shuttle's ceiling as much as she could.

The crew crossed off the major items on their to-do list one by

one. First up on Day One was deploying the first of their two satellites. Working with John, Sally flipped all the necessary switches
and pulled all the right levers to release a Canadian communications
satellite called Anik C2 into the void. The next day saw the deployment of an Indonesian communications satellite called Palapa B1.
Both satellites, giant dark cylinders the size of a bus, twirled out
into space just as intended.

With those deployments out of the way, it was time to have
some fun.

The prime reason Sally was on this flight was to perform the
SPAS-01 demonstration, using her deft robotic-arm skills to maneuver an experimental payload out into space. On the fourth day,
she and John positioned themselves in the cockpit, eager to show
off their training. Stationed at the arm's controls, John took the
first shift, extending the arm out of the open payload bay and
manipulating it into a position high above the cargo hold, like an
arcade claw poised to snag its prize. Certain of his accuracy, he then
lowered the arm and plucked SPAS-01 out of its cradle, pulling it
up and out of the bay.

As if placing a book on a shelf, John dragged the arm out into
space and released SPAS-01—the payload remaining seemingly
motionless in its new position. After the payload was freed, it was
time to perform their planned orbital dance. Crip piloted the Shuttle down and away from the satellite, rotating the orbiter so that
its open payload bay faced the SPAS-01 flying free overhead. That
moment was ripe for a photo opp. Sally issued a command on *Challenger*'s computer, prompting SPAS-01 to use its onboard camera to
capture the Shuttle. The lens snapped a dark, moody photograph
of *Challenger*, perched against the backdrop of a royal blue ocean

and a sprinkling of snowy-white clouds. Content with their work, Crip flew back to the payload and John moved the arm up and out again, grappling the payload once more. Worried about the SPAS-01 heating up, they moved it into the Shuttle's shadow to keep it cool.

After lunch, it was Sally's turn.

Mimicking what John had done in the morning, Sally steered the arm with its prize in tow, again dragging the SPAS-01 payload out into space before leaving it on its own. Sally controlled the arm, guiding it out of the way of the satellite so that Rick, this time in charge of the controls, piloted *Challenger* out and around the satellite, practicing a different kind of distancing maneuver than Crip had done.

In the middle of their demonstrations, Sally and John didn't forget to capture the images of the robotic arm they'd practiced so hard to snag. At one point, with SPAS-01 floating in front of the Shuttle, they controlled the arm, bending its joints in just the right way so that it resembled the number 7. They commanded SPAS-01 to photograph a clear shot of the arm sticking high out of the payload bay, looking like a 7 for STS-7.

"Those pictures are a very important part of my memory of the flight," John said years later. Beaming with pride at their work, the crew eventually had to tuck SPAS-01 back into its cradle. One last time, Sally controlled the arm back up to the satellite and hooked on it once more. As she made her crucial maneuver, uncertainty gripped her. *This is real metal that will hit real metal if I miss,* she thought. *What if we don't capture this satellite?*

But she'd trained hundreds of hours for this exact moment. And she didn't miss. Just as she and John had rehearsed, she neatly stowed SPAS-01 back into the payload bay, their biggest job complete.

All in all, it was a fairly flawless mission, save for one scary moment midflight. While performing some procedure at the control panel, Rick turned and noticed a small crack in the Space Shuttle's windshield. A small particle had penetrated partly into the windowpane, creating a tiny impact crater. Fortunately, the projectile didn't pass all the way through. If the hole had been any bigger or deeper, the crew would have been killed instantly. Overall, the window was deemed fine by the crew, and Crip decided not to tell Mission Control, figuring it would just worry the engineers needlessly. (He'd later get in a spot of trouble for that decision.) They all simply decided to keep an eye on it. Later, they'd learn that the crater was caused by a tiny fleck of paint. The smallest of objects can have an outsize impact when moving at 17,500 miles per hour.

During the daytime hours, Sally was all work. Though for her, the work was pure fun. But during the nighttime or in the early-morning hours just after waking up, the astronauts would take turns just staring out the Shuttle's windows. A couple days into the flight, Sally got out her Sony Walkman and began playing the mixtapes she'd crafted for the trip. With the sounds of the Beatles, the Beach Boys, or Janis Joplin pulsing through her headphones, she'd simply stare at the planet's luminous glow down below.

"I never got tired of looking out the window at Earth," Sally said later. "It's just a constantly changing view, and it's just a beautiful sight."

One moment coral reefs peeked out at her from under the turquoise waters surrounding Australia. The next, a massive orange dust storm emerged over Northern Africa, crawling slowly across the continent. Gargantuan white cyclones swirled over the oceans, looking like puffy discs bent on destruction. "It was just a spectacular feeling to be able to look down at the moon's reflection along rivers and watch our progress up the East Coast of the US," she said.

But what stood out to Sally most wasn't any one geographic feature. It was Earth's atmosphere, standing in stark contrast against the blackness of space. "It looked as if someone had taken a royal blue crayon and just traced along Earth's horizon," she said. "And then I realized that that blue line—that really thin royal blue line—was Earth's atmosphere. And that was all there was of it."

In those moments looking out the window, Sally realized just how fragile human life is. From nearly two hundred miles up, the air that keeps all of humanity alive looks like nothing more than a fuzzy thin band surrounding Earth's outer edges. The sight gave her renewed resolve to ensure that this atmosphere stayed intact for generations to come.

———

"CHALLENGER, THIS IS Houston," astronaut Mary Cleave called up from the ground in Mission Control. "How do you read?" Mary was one of NASA's newest astronauts, who'd been selected in a fresh batch of recruits picked in 1980. The latest group had included another woman, too: Bonnie Dunbar, who'd been an engineer at JSC before her selection. Now, with their addition, there were eight women in the astronaut corps.

"I can hear you, Mary," Sally responded from space.

"Good evening, Sally. Sorry to wake you up," Mary said. She proceeded to give Sally instructions for making a small adjustment to the Shuttle's onboard computer. It was a completely normal conversation—fairly boring to listen to—but the substance wasn't what mattered. It was the first time that a woman on Earth had spoken to a woman in space. Mary and Sally didn't even recognize the moment at all. They were just having a conversation. A reporter later lamented to Mary how "disappointing" the conversation was for such a special occasion.

But that was how Sally operated. For the six days that she was in space she simply did her job, trying her best to ignore any history that was being made. She couldn't help it that STS-7 was simply bursting with historic "firsts." Even its landing was supposed to make a statement. Their mission was slated to perform the first Shuttle landing in Florida. All previous Shuttle missions had landed at Edwards Air Force Base in California. But NASA had been working on a fancy new runway at Kennedy Space Center, and the concrete was finally ready to feel the Shuttle's massive tires touch down.

On the day of reentry, however, that historic plan was thwarted. A thick fog bank had rolled over the landing strip, completely obscuring the landscape. Slightly bummed, the astronauts made plans to divert to Edwards. After cleaning their equipment and strapping back into their seats, Crip and Rick steered the Shuttle out of orbit, sending the vehicle on its dive toward Earth. Out the window, Sally watched the bright glow of the atmospheric plasma, heating up and enveloping the Shuttle as it sliced through the thick air surrounding Earth.

For Sally, the detour turned out to be a blessing in disguise. As she walked down the stairs leading out of the crew cabin and onto solid ground, only a small crowd of air force personnel and their families were on hand to greet her and the incoming crew. She was 2,500 miles from the thousands who'd gathered at Cape Canaveral, all clamoring to see her arrive. After a quick phone call with President Reagan, who joked that the crew had forgotten to pick him up in D.C. on their way home, Sally addressed the crowd out in the California heat.

"The thing that I'll remember most about that flight is that it was fun," Sally told the crowd. "And in fact, I'm sure it's the most fun I'll ever have in my life."

THAT FUN THAT Sally had in orbit would soon be eclipsed by what was waiting in her new reality.

Flying home on a NASA jet, she and the rest of the crew arrived in Houston, and her husband, Steve, was there on the runway to see her. With a big smile, Sally wrapped her arms around him in a tight hug, happy to be reunited. After the rest of the crew had embraced their spouses, they all made their way to limousines that would take them to JSC for one final presentation to an adoring crowd. While at the airport, someone handed Sally a giant bouquet of white roses, adorned with an ostentatious cream bow. She wasn't the only one to get flowers, either. All the wives of the male crewmates had each been given a red rose, though Steve hadn't been given anything. Sally carried the massive bouquet with her until they all reached JSC.

It was then that Rosegate occurred.

A large crowd had gathered on the JSC campus, awaiting the STS-7 crew. As the five crewmates and their spouses stepped out of Building 1 to address this latest group of fans, Sally quickly turned and gave the bouquet to a NASA protocol officer. The motive was plain: she just wanted her hands free. Without the bouquet, she stood with her arm around Steve, before speaking to the crowd who cheered her on. Once the presentation was over, the officer who'd taken the flowers returned, trying to give them back. Sally, with Steve next to her, declined to take them, instead turning to talk with George Abbey.

She had no idea the controversy she just ignited.

The next day, a newspaper writer interpreted the act as some feminist statement—a way for Sally to establish equality with her male crewmates. Soon afterward, the letters began pouring in. She'd later write, "That one little action—giving back the flowers—probably touched off more mail to me than anything I ever did or said as an astronaut."

That moment was a harbinger of what was to come. All eyes were on Sally now. Back in the gravity-filled environment of Earth and without a mission to train for, she was now almost completely exposed to the adoring public who wanted to witness her every move. It began the day she came back to Houston, as crowds of neighbors and news crews gathered outside her and Steve's home, tying yellow ribbons around the trees in her yard and holding up a banner that said, "WELCOME SALLY!"

But Sally and Steve had long planned to escape their abode that night by booking a nearby hotel room. That plan fizzled when Steve went to check in and spotted a reporter he knew hanging around. Instead, the couple went to Dan Brandenstein's house to spend the night, leading the press to note Sally's absence at her own home.

It was only a small glimpse at the road ahead.

In just the first month back, Sally and her fellow crew members found themselves speeding through eight different states. They traveled to New York, where they received a key to the city from Mayor Ed Koch. They attended an opulent reception at the National Air and Space Museum with five hundred attendees, plenty of whom asked for Sally's picture and autograph. There was an intricate military ball at the Dunes Hotel & Country Club in Las Vegas. Again, Sally and the STS-7 crew found themselves at the White House, dining with President Reagan. Sally sat between Reagan and the prime minister of Bahrain, with whom she discussed the feeling of being weightless.

And the whoosh of attention showed no signs of abating. Requests were streaming into NASA to book Sally for every minuscule event. Less than a week after she returned from her flight, the agency received more than a thousand media appearance requests for her. At one point, calls into NASA's press office asking for Sally peaked at twenty-three an hour.

With the shield of training gone, Sally's crewmates tried to become her new shield. They made an effort to go with her to as many events as possible—to serve as a buffer between their colleague and the press. "We would minimize those appearances before the flight as much as possible," Crip said. "And we did that. But she still had a lot of it, and after the flight, some of that protection wasn't there." Norm Thagard unfortunately served as a literal buffer during one press event in D.C., when hordes of TV news crews shoved him against the wall as they clamored to get the perfect shot of Sally.

Worse, the requests were starting to get bizarre, like the artist who wanted Sally to sit for a portrait created out of jellybeans. Or the production company that wanted her to act in a comedy where a child dreamt he met Sally Ride while traveling in space. As the requests continued to mount, Sally started to feel more and more out of control of her life. She was an introvert at heart, not one to seek the spotlight willingly. And here she was, seemingly meeting everyone on the planet.

One day, NASA received an invitation for Sally to visit the newly minted Sally K. Ride Elementary School located an hour north of JSC in Conroe, Texas. When Sally received word, she made it clear she did *not* want to go. But the school desperately wanted her, even calling the director of JSC to secure Sally's booking. Realizing she couldn't get out of it, she made a demand that she'd only go if Carolyn Huntoon came with her.

So the two embarked on the short road trip north, getting slightly lost along the way. Once they finally found the school, they drove up and made their way in, only to start hearing a children's choir start singing as they walked through the front doors.

"We are proud of our school, Sally K. Ride! We will always take a challenge and always do our best . . ."

Panicking, Sally turned to Carolyn. "I don't think I can do this."

"Yes, you can, go on," Carolyn replied. She took Sally by the arm and walked her into the school, where she was reluctantly showered with gifts and subjected to a few more verses of the school's new song.

As the requests mounted, Sally pulled away more and more. It all came to a head when she got a request straight from NASA Headquarters' public affairs to go on a new Bob Hope special. The popular comedian was hosting a special tribute to NASA called *Bob Hope's Salute to NASA: 25 Years of Reaching for the Stars*, and some famous astronauts including Neil Armstrong had agreed to be interviewed. The production team also wanted Sally. NASA was told that this wouldn't be a regular comedy show, but a serious discussion with Sally about her thoughts on the world and what her time was like in space. It didn't matter. Sally turned it down.

Reluctant to take no for an answer, headquarters enlisted JSC director Gerry Griffin to try to persuade Sally to change her mind.

He found her at work, sat her down, and asked if she'd heard about the request. She said that she had. And she wasn't going to do it.

"Why not?" he asked.

"Because he's a sexist," Sally replied. Sally brought up Bob Hope's USO tour during World War II, which involved parading women in scantily clad attire to entertain the troops. It didn't matter that she was promised a "serious" show—she didn't like this man's reputation.

Gerry continued pressing but Sally stood firm. Soon after that, she just vanished.

She didn't tell anyone where she was going, not even Steve. But that wasn't particularly abnormal for her. She'd done this disappearing act before. Steve understood her desire to escape, though he didn't think it was particularly responsible in these circumstances. Finally, after a week of absence, she called Steve to tell him she was okay and that she had skipped off to California, as he had suspected. She'd taken refuge with Molly and Molly's partner in Menlo Park, making sure that an appearance on the Bob Hope special wouldn't be imposed on her.

It was beginning to dawn on Sally that she needed some help. All her life she'd been a happy individual, excited to wake up and start each day. Now, she realized she wasn't that same happy person. She woke up nervous, filled with anxiety about what each day might bring.

"Swarms of people surrounding her and people wanting to touch her, take her photograph, invite her to things—she slowly realized it was really getting to her," Tam said.

In a moment of clarity, she turned to therapy. It's unclear who exactly she spoke with, though Steve heard later that she supposedly talked to Terry McGuire, the good cop psychiatrist whom they'd all met during the selection process. Either way, Sally knew she needed an empathetic listener who could help her through this unusual time.

For the most part, Sally viewed her press odyssey as a nightmare. But scattered amid the dizzying array of speeches and networking events, there were special moments she came to cherish. She made multiple appearances on *Sesame Street*, where she loved meeting with the science-curious children. They asked the fun questions, such as what it was like to go to the bathroom in space. And of course, she got to meet some of the people she most idolized. Through her fame, she met Betty Friedan and reconnected with Billie Jean King, whom she'd hit balls with as a young tennis player.

But perhaps the most intriguing person Sally met was one she hadn't expected to *ever* meet. At a reception in Hungary where Sally and Steve were attending a meeting of the International Astronautical Federation, Sally felt a tap on her elbow. She turned to find Svetlana Savitskaya, the cosmonaut who'd become the second Soviet woman to fly to space in 1982.

"Sally," Svetlana acknowledged.

"Hello," Sally said, cautious.

"Congratulations on your flight."

"Congratulations to you, too," Sally replied.

Their conversation didn't last long—relations between the US and the Soviet Union had been exceedingly chilly in prior months— but later that day Sally heard Svetlana talk and decided she was a genuinely good person. Somewhat surreptitiously, Sally contrived to arrange a more confidential meeting.

Using as an intermediary a Hungarian physicist she knew, she expressed an interest in meeting Svetlana again. Thanks to some behind-the-scenes machinations, Sally wound up being invited to meet a small group of Soviet cosmonauts at the apartment of Hungarian cosmonaut Bertalan Farkas. Steve disapproved of the meetup, conscious of NASA's dim view of American astronauts mingling with Soviet cosmonauts, so he hung back at a coffee shop. Sally, though, was determined.

The mood at first was tense, with neither party knowing quite how to comport themselves. Sally tried smiling to put everyone at ease. The cosmonauts immediately made a joke about "no press," which eased everyone's anxieties. Svetlana then walked over to Sally and sat in a big armchair, right next to hers.

From that moment, the women instantly connected. Svetlana peppered Sally with questions, asking her how long she'd trained and about her flying experience. They swapped stories about the respective spacecraft they flew, with Svetlana fascinated by how the Space Shuttle landed on a runway. At one point, they all devolved into laughter about how they slept in space, with one of the cosmonauts demonstrating by floating his arms into the air.

As they spoke, Sally found that she really enjoyed talking with Svetlana. "I felt closer to her than I felt to anyone in a very long time," she said. "And it was partly just that I understood a lot of what she had been through."

Sally thought to herself that Svetlana would have easily made the astronaut corps in the US. In fact, she might have even beaten Sally in the race to be the first American woman to orbit if the two had been pitted against each other, she thought. Ultimately, Sally saw a lot of similarities between her and Svetlana, thinking that the cosmonaut reminded her of her colleague Shannon most of all.

Sally enjoyed her time at the gathering so much that she stayed for six hours. Long after midnight, she and Svetlana exchanged a few final words and hugged goodbye before their staggered departures.

"I left with the feeling that we would probably meet again," Sally said. "And if we did, we would be just as close as we were at that moment."

The whirlwind press tour had brought unbelievable highs and remarkable lows. In the middle of it all, Sally sat down with Gloria Steinem, agreeing to an interview to talk about the surreal few months she'd been experiencing. Sally discussed all aspects of her flight, placing particular emphasis on the SPAS-01 demonstration and the pictures they were able to take while in space.

"That's probably what our flight will be remembered for, I think, is those pictures," Sally said.

"Want to bet?" Gloria replied.

CHAPTER 12

Take Two

"Would you have liked to have been the first woman in space?"

Judy paused briefly, considering how to answer the reporter's question. It was a query she'd been getting all the time since being assigned to her flight in early 1983. The press had a one-track mind, and all they could think about was the crew lineup and each person's place in it. Surely, Judy must've been upset about being in "second place" behind Sally, they thought. She took a moment before replying into the reporter's microphone while a camera pointed at her face.

"I'd like to be *any . . . woman* in space, thank you," she said with a smile.

It was the perfect answer, but it was also true. Judy just wanted to fly. And the reality was she was perfectly fine with her place in line. She hadn't made her emotions about Sally's selection well known, but even before the decision had been made, she'd told a reporter that it wasn't at the top of her mind. "I'm not here to get my name in the books or to make a name as the first woman astronaut," she said. "I'm here to contribute to the space program and do the best job that I can." However, her father cast some doubt on that, once telling a paper that Judy was "somewhat disappointed but not *really* disappointed" about not being first.

What is certainly true is that Judy witnessed firsthand the avalanche of requests that came crashing down on Sally after the end of STS-7, and to Judy, it didn't look like fun. Both Sally and Judy commiserated over their shared aversion to the press, so Judy knew the attention must've been hard for Sally to endure. Every now and then, Sally charged into Judy's office and slumped down in an armchair, exhausted at all the requests she was getting. Any disappointment Judy had felt at not being the first woman to fly quickly evaporated as she watched Sally's PR odyssey, and Judy confided in Sally's husband, Steve, one day that she was quite content to go second.

Judy learned of her assignment in early February 1983, before Sally had even flown. She received word one morning that George Abbey wanted to see her. When she walked over to his office on the eighth floor of Building 1, she found herself standing with three of her TFNG classmates—Mike Coats, Steve Hawley, and Mike Mullane—as well as Henry "Hank" Hartsfield, an older astronaut who'd flown on the fourth flight of the Space Shuttle. The five stood together, wondering silently if this could finally be it for them. Could this be an actual flight crew?

A secretary ushered them into George's office, where they all sat around a table. "Well, you know, been thinking it's time to give you guys a new assignment," he said. In his signature low, soft-spoken style, George asked them if they'd be interested in flying on an upcoming mission called STS-12. This wouldn't be any ordinary flight. It'd be the first flight of the space shuttle *Discovery*, a brand-new vehicle that was currently being built by Rockwell International out in California.

Though their flight started off as STS-12, its title soon morphed into STS-41-D. That fall, NASA decided to change the naming system for the flights—to *avoid* confusion, they were told. The payload that was supposed to fly on STS-10 had been delayed, which would mean STS-11 would be flying first. NASA thought it would be confusing, though, for the public to mentally process STS-11 flying before STS-10, so the agency came up with a wacky numbering scheme for the missions moving forward. The "4" in STS-41-D referred to the flight being set to launch in 1984. The "1" referred to its planned launch from Cape Canaveral, the first of two possible launch sites (the other being Vandenberg Air Force Base in California). And the "D" referred to the mission's payload being the fourth one manifested for fiscal year 1984.

This was supposedly what a less complicated name looked like, in NASA's eyes. Later, the astronauts would speculate there was a much more superstitious reason behind the name changes: NASA simply didn't want to have an STS-13, given how Apollo 13 had fared.

To Judy and the others whom George sat down with that day, they couldn't care less what the flight was called. All that mattered was that they were now in the crew rotation. And Judy already knew some of her crewmates well. She'd worked with Steve and the Mikes before, and she'd grown closer to Steve because of her friendship with Sally. Mike Mullane had a reputation as a prankster who didn't necessarily pride himself on being politically correct, and his mouth would often get him into trouble. But Judy knew how to deal with that type. She didn't take crap from anyone and could easily throw it right back when it was hurled her way.

Sworn to secrecy, Judy and Steve went out to dinner that night to celebrate their selection. Together they talked about what training might be like, based on their respective conversations with Sally. Judy also just felt unbelievably grateful that Sally would be paving the way for her. "I think Sally did an outstanding job of fielding a difficult situation," Judy told a journalist later. "A very bright spotlight was on her as the first one to go, and she handled it very well." Fewer eyes on Judy also meant less pressure to not screw up—and perhaps a little more room to have fun.

And fun wasn't going to be hard to have with this crew. The STS-41-D team quickly established itself as a rowdy bunch. They weren't 41-D, but the Zoo Crew, a nickname that the astronaut corps would bestow on them over time. The impetus for the name came from a trip that Mullane and Steve took to the Seychelles, of all places, off the coast of East Africa. The US Air Force had a ground control station there, a facility that could relay data to and from the Space Shuttle as the vehicle flew overhead. George had wanted the astronauts to talk to the workforces at the various ground control stations around the world, and Mike and Steve couldn't believe their luck when they found themselves enjoying the white sandy beaches and ritzy resorts of the Indian Ocean island chain. They had *so* much fun, in fact, that they wound up missing their flight home, forcing them to stay two more days. While running down the clock on their extended "vacation," Steve and Mike inadvertently ran into Hollywood superstar Bo Derek and her husband, John, who were scouting locations for their next movie, *Tarzan*. Mike proudly identified himself and Steve as astronauts and begged to have a photo taken with them.

Still starstruck from the run-in, Mike told everyone about the celebrity sighting when he and Steve returned to work. Inspired by his story, Judy decided to give the guys nicknames. Mullane became "Tarzan" while Steve became "Cheetah." Eventually, everyone in the crew got in on it. Mike Coats, with his square-cut jaw and award-winning smile, became Superman, while Hank became the Zookeeper, or Zeke. Naturally, Judy, the lone woman of the group, became Jane to Mullane's Tarzan. Though that was actually her second nickname at NASA. To everyone she knew, Judy wasn't actually Judy. She was J.R. But now she answered to multiple monikers. When Mike Mullane gave her the nickname, he joked: "Would you like to swing on my vine?"

"Sure, Tarzan. But first I'll have to tie a knot in it so I have something to hold on to," Judy handed him back.

The Zoo Crew proceeded to live up to their moniker. They were always playing pranks and cracking jokes. Knowing that Mike Mullane hated sports, Steve and Mike Coats would often quiz him on sports trivia questions. Mullane would always pretend to get angry and reply with "Earl Campbell," a Houston Oilers running back and the only professional athlete he knew. To honor their commander Hank's fiftieth birthday, the crew, who were mindful of his conservative political leanings, gave him an extra-special gift: a copy of *Ms.* magazine signed by Gloria Steinem. And Judy's crewmates repeatedly teased her about her big crush on the actor Tom Selleck, whose picture she liked to pin on the walls of her workplace.

One workday, someone asked the crew if they wanted a soda from the vending machine. Judy, who hailed from the Midwest, replied, "Yeah, I'll have a pop." Soon the guys all started to razz her for her colloquialism. "Pop? I said I'm going to get soda," one cried.

It eventually flared into a back-and-forth debate over the proper way to describe carbonated beverages.

As the crew's only female astronaut—and with a reputation for going easy on the guys—Judy found herself the butt of many jokes from Mullane. One time, she walked into the building that housed the robotic-arm mockup, which mission specialists used for training. Mullane had been operating it just before she came in, and as she walked toward the training cockpit, he maneuvered the tip of the spindly arm down low, facing her feet. The end of the arm then tracked her as she walked, moving up from her legs to her face. Knowing that the robotic arm had a camera at the end, Judy climbed into the cockpit, seething. "You're a pig, Mullane," she told him.

"Judy was a target because she took those jokes pretty well," Mullane would recall years later.

She never really got mad, not seriously, anyway. Just like Sally, Judy didn't want to make a big deal of her gender. She didn't want people to think that she'd gotten to her position based on some predetermined trait that was out of her control. She felt that she'd worked unbelievably hard to get to the level she was at. "I think I'm where I am because I just happened to make the right decisions at the times when the decisions were presented to me," she told a reporter once. When asked if she credited the women's movement with her success, Judy said she didn't credit anyone.

However, she was making history, whether she wanted to or not. With her flight, she'd become the first Jewish American in space. It was another "first" the press clamored to ask her about, and one that Judy didn't really want to highlight either. One Jewish publication approached her to talk about her faith, but Judy ultimately declined to make a big fuss about it. The article made it seem as if Judy had downplayed her connection to Judaism, but the truth was she just didn't like being labeled. She once told her father: "Dad, I don't want to be a Jewish astronaut. I don't want to be a Jewish woman astronaut. I just want to be an astronaut, period. I just want to go out in space and do my job."

As much as Judy tried to avoid drawing attention to herself, it wasn't easy. She was objectively a knockout. She'd also gotten into even better shape during her time at JSC, having picked up frequent jogging like the rest of the astronauts. And that could attract unwanted notice from the men she interacted with. Once, during a trip to Los Angeles to meet with some Shuttle contractors, Judy found herself being followed by an overeager engineer while on a factory tour. Each time she turned her head, he always seemed to be by her side, asking if she wanted water or a snack. She tried to tell him no as politely as possible, but he just wouldn't take the hint. He'd open every door for her and then run to the next to be there when she arrived. It was exactly the kind of singling out she despised, but she stayed as civil as possible.

She thought she'd seen the last of him when she found herself in the cockpit of a T-38 to return home that evening. But to her astonishment, the same man strode out to the landing strip and pulled out the blocks from beneath her plane's wings. It wouldn't be the last weird place he popped up. Soon she started getting creepy letters and poems from him at her house. Then came a package that mysteriously wasn't postmarked. Even with JSC security on high alert, the engineer showed up at the office one day and beelined to Judy's desk, asking for an autograph and for her to be his pen pal. Thankfully, Mullane was nearby and made a hasty excuse to pull her into a "meeting." After that, NASA management did *something* that halted the employee's repeated harassment.

An overzealous stalker was the last thing Judy wanted to deal with. Her job was her prime focus. She'd thrown herself into her work, something her friends and family back home easily noticed. "There's no question now, she's married to her career," her father, Marvin, told a reporter, noting that Judy didn't visit Ohio as much those days.

Judy's focus may have been steadfast, but her jobs would shift for this mission. While Sally's priorities had stayed pretty much the same for STS-7, the STS-41-D team saw their flight evolve over time. "Don't fall in love with your payload," was a common adage among those in the Astronaut Office. And the phrase applied to 41-D. Originally, the crew had trained to launch a massive communications satellite for NASA called TDRS (pronounced Tee-Druss), one that would help relay data from a spacecraft in orbit to Earth. But on an earlier flight, engineers found a problem with the booster needed to propel the TDRS. So, it got pulled from 41-D. While the crew sweated over what the changes meant for them, their luck turned. They were given a pair of smaller communications satellites to launch instead.

With the shuffle, Judy caught an assignment all her own. Once the crew deployed their main satellites, Judy would be responsible for unfurling a long rectangular solar array. She'd be the one to flip the appropriate switches in the cockpit, rolling the array from inside the Shuttle's cargo bay out into space, where it'd straighten and stiffen like a board. NASA hoped to test the durability of this elongated solar panel, to see if it could hold up in the extreme space environment. If so, the design could someday be used to help future spacecraft collect solar power. Judy's expertise with electrical engineering made her the perfect person to handle the array, even though this test was more about the array's resilience than collecting energy. And she'd also need to stay up-to-date on her robotic-arm skills. The Canadian appendage would be on board in case the solar array got stuck in space; in that case Judy would use the arm to tuck the thing back in.

Judy also volunteered for another assignment: working backup to a new crewmate who'd been added late in the game. Charlie Walker was assigned to the crew about five months in, but he wasn't a typical crewmate. Technically, he wasn't even an astronaut. He worked for a private contractor, McDonnell Douglas, making him the first payload specialist to climb aboard the Space Shuttle. His job on board was to perform an experiment for his company to see if pharmaceuticals could be purified better in a microgravity environment.

Unlike Sally, Judy wouldn't have much of a view at launch. She'd be sitting down in the mid-deck with Charlie, the room beneath the cockpit that had just one small window and practically no view of the outside world. Judy and Mullane had flipped a coin for it. She'd fly in the mid-deck during ascent while he'd sit there during reentry. That meant during the climb to space she'd be staring ahead at a wall of lockers, with no frame of reference for their position in the sky, while the four crew members on the top level would get front-row seats to the action. She told them they weren't allowed to say anything vague when they made the climb to orbit, like, "Did you see *that*?" "What was *that*?" She threatened them with pain to be as specific as possible in their observations.

Because she wouldn't be sitting in the main cockpit, Judy was usually off the hook whenever the crew did simulations to train for ascent and reentry. The motion-based simulator only had room for the top four seats, and there wasn't much Judy could do from her sequestered chair in the mid-deck. But that didn't stop her from attending the sims. She wanted to consume as much as she could about the training process, and she'd often sit at the console with the training instructors, just listening to the phrases uttered during ascent.

It was the best way for her to learn. One day, she might be up there on the top deck, and she wanted to be as prepared as possible.

As the launch approached, the issue of family was becoming hard to avoid. Judy knew she needed to send out invitations to the launch, but it felt as if a black cloud hung over the task. When she'd sent out invitations for her wedding to Michael Oldak, two separate sets had been mailed: one to her father's side and one to her mother's. Now there was pressure to invite the entire Resnik clan to Cape Canaveral, including Judy's mother, which she was reluctant to do. At one point, the tension boiled over in a call with her brother, which Mullane overheard at work.

In the end, everyone wound up getting an invitation, perhaps because it was the path of least resistance. Her father and brother got them. Her stepmother and stepsisters got them. Even her mother got one. Plus, the men of her past were coming, too. Michael gladly accepted, having become a good friend since their divorce. And so did Len Nahmi, whom Judy still had a connection with, even if it was on and off.

In the weeks before the launch, Judy and Mullane traveled to Florida—a quick trip to test out some payloads ahead of the flight. They arrived the day before their work was set to begin, and when they landed, they jumped in a convertible bound for Kennedy Space

Center. There, they toured the old launch facilities, sidestepping any hidden security guards, and Judy quizzed Mullane on his rocket history knowledge. Afterward, the pair drove off to the astronaut beach house for a beer. They sat together in the sand, looking out at the waves lapping against the shore, and toasted being the next crew on deck.

In that moment, Judy opened up to Mike about her family. She liked Mullane; he and his wife had often welcomed her into their home, as had Hank and his family. Such hospitality wasn't shared by plenty of the other astronauts' wives, who felt threatened by Judy. To her friend that day, she spilled the intimate details of her personal life, telling him of her struggles with her mom and how hard it had been to grow up under her restrictive rules. It was a rare moment of vulnerability that she didn't like to show too often. Mike simply listened.

After the conversation settled, the two left to prepare for their busy day ahead. And soon, Judy was far too consumed with launch preparations to worry about family affairs.

"WELL, GANG, WE'RE going to go now unless something really bad happens," Hank said to the crew.

Believe it or not, those were meant to be words of comfort for Judy and the team as they sat horizontal in their metal chairs aboard the space shuttle *Discovery* on June 26. They had just thirty seconds to go in the countdown, and it was finally feeling like space was within their grasp. T-minus 30 seconds was also much closer to liftoff than they had gotten the day before, when they were *supposed* to launch.

It had been an agonizing abort. Judy had woken up early, gotten dressed, and eaten breakfast, believing she'd finally be an astronaut at last. She'd made it through every critical step, just as Sally had: walking out of the crew quarters, driving out to the launchpad, looking out over the gantry, before strapping into her seat in the mid-deck. With every milestone, the tension heightened and the anxiety built. But at T-minus 32 minutes, flight controllers informed the crew there was a problem with the backup flight computer. They entered a planned hold of the countdown at T-minus nine minutes, when the clock was stopped intentionally. But whatever the issue was couldn't be fixed on time, and they never came out of the hold. "*Discovery*, we're going to have to pull you out and try again tomorrow," a flight controller said over the intercom.

Every emotion she'd felt the day before, she replayed again the next morning as she suited up and journeyed out to the pad with her crew. But this time, it was starting to feel even more real. They successfully made it through T-minus 32 minutes with no glitches, and when they entered the T-minus 20 minute hold, the countdown picked up again as planned, and the backup computer started to run. The minutes disappeared one by one, until they were just thirty seconds out, prompting Hank to make his proclamation. By that point, the Shuttle computer had taken over. It felt like nothing could stand in their way.

Judy turned to Charlie lying next to her, who looked over at her. Both their visors were closed by that point, obscuring their faces. But they extended their arms toward each other and gripped each other's hands for some last-minute solidarity. At that moment, Judy was glad to have Charlie with her down below.

She then looked ahead and listened to the countdown, her

pulse quickening. "We have a go for main engine start," the announcer said on NASA TV. "Seven. Six. Five. We have main engine start."

A deafening blast sounded. The engines ignited at the Shuttle's base, and *Discovery* started to tremble violently, sending vibrations through Judy's entire body. The massive vehicle began to teeter backward—that twang they'd all been expecting. Judy braced herself for the moment *Discovery* would lurch off the launchpad. It was the moment she'd been waiting for for years.

But just a second after the shaking had started, the movement suddenly ceased. A dreadful mechanical wrenching sounded throughout the cabin, and the engines, roaring just a moment ago, became silent. An eerie stillness filled the air, only to be broken by the Shuttle's master alarm, which started wailing like a police siren throughout the vehicle. *Discovery* sat on its launchpad, motionless.

That . . . wasn't supposed to happen.

"We have a cutoff, and we have an abort by the onboard computers of the orbiter *Discovery*," the announcer said to the public over the loudspeakers.

Everyone on the launchpad stayed absolutely silent, save for the seagulls outside screaming at the loud disruption. Judy's mind began to race. The engines had just shut off inexplicably, an instant after igniting. A pad abort had never happened before. Once she wrapped her head around what was happening, she snapped into focus. It was possible that the astronauts would need to make a break for it in case the Shuttle was unstable. And it was her responsibility to open the hatch door to let everyone out. She'd practiced this exact moment during training, but she didn't actually expect to ever use the expertise. As her heart thumped, she stared ahead of her, trying to remember which knob to turn on the hatch door and how fast she could get the thing to open.

No one gave her the green light to open the door, though. Everyone stayed ghostly quiet, a fairly abnormal scene for the talkative crew. Right away, Launch Control declared that there'd been an RSLS abort, or a redundant set launch sequencer. It was a fancy term that meant the computers had picked up some unusual readings and shut down the main engines before the solid rocket boosters could ignite. Since the boosters weren't currently tearing apart the Shuttle's hull, everyone figured they hadn't fired. Judy and the rest of the crew said silent prayers for the solids to stay off. If those boosters *did* ignite, it would mean instant death for all of them.

Over the intercom, Judy could hear Hank and Mike Coats going back and forth with Mission Control about what was going on. Suddenly, the flight controllers began to panic. "Break, break, break, break!" a controller yelled into the intercom. They urged Mike to shut down one of the engines—fast. Their data showed that one of the three was still on. The crew knew that couldn't be right. The Space Shuttle was just sitting there. Looking at the control panel, Hank and Mike saw two red engine lights illuminated, indicating that two of the engines were in shutdown mode. The third engine light, however, was off. That freaked out Mission Control, leading them to believe the engine was still running. The truth was, the engine never ignited at all. They'd all learn later that there'd been a blockage in one of the engine's hydrogen valves, which prevented it from starting and ultimately triggered the abort. But during those first confusing moments, the flight controllers didn't know what was going on. Eventually, everyone realized that no engines were turning on anytime soon.

Amid that confusion, the tension in the cabin grew thick. No one spoke unless they were responding to personnel on the ground. Everyone had questions, but they didn't want to interrupt the teams frantically working on the issue. Judy just listened, waiting for instructions and hoping that everything would be okay. It was the first time in Shuttle history that the vehicle's engines shut down right before liftoff. No one knew what to expect.

After a few minutes of quiet panic, mission controllers informed Hank that everything was safe. Then Steve piped up and broke the silence.

"Gee, I thought we'd be a lot higher at MECO," he joked.

That did it. The tension snapped, and the anxious crew let out some much-needed laughter, while Mike Mullane cursed out his friend in the next chair. Steve's joke gave them all an invitation to banter a bit, but it wasn't long until the stress returned. Flight controllers said the word no one wanted to hear: *fire*. Flames had been detected on the left side of the vehicle. All of a sudden, valves of water surrounding the Shuttle's exterior burst open, spraying thousands of gallons of liquid all around the launchpad to quench the flames. Just hearing the word *fire* sent chills through everyone inside *Discovery*. A blaze near a rocket loaded with more than 500,000 gallons of extremely combustible liquid propellants was a scary recipe for unintended explosions.

Hank told everyone to unstrap and get ready to move. If they bailed out, it would have to be quick. They'd all have to run across the catwalk and nearby launch tower toward a series of metal baskets, which would take them to the ground via suspended cables—the world's most intense and scariest zip line. But they remained in their seats, waiting for directions. Finally, Judy couldn't take it much longer. Breaking from the rest, she got out of her seat and made her way over to the hatch, peering out the lone window on the mid-deck. She could see the water from the fire suppression system dripping down the side of the Shuttle, like torrential rain cascading down a windowpane. The suspended walkway had also snapped into position outside the door, providing them an escape route to the baskets if needed. Judy couldn't see any fire, though, at least not from where she stood.

"Do you want me to open the hatch?" she asked Hank.

Hank stalled on the question. They all debated whether it was wiser to stay seated or to make their grand escape. The last place they wanted to be was *inside* the Shuttle if the fire made its way to the external tank. Eventually, he made a judgment call, telling Judy no. They were going to wait it out.

After a while, things calmed down and, about forty minutes after the abort, flight personnel arrived to open the hatch and help the crew out of the Shuttle. Judy was the first to emerge, smiling and happy to be free from the cabin. As she walked across the suspended walkway, the residual water from the fire suppression system dripped from the lattices above her, soaking her and her flight suit. When they stepped inside the elevator, an inch of water awaited them on the floor.

Down at the launchpad's base, the crew gathered inside the designated crew transport vehicle known as the Astrovan, which had been upgraded to a modified Airstream motorhome prior to the flight. With the air-conditioning on full blast, the six drenched crew members shivered as they rode back to their quarters. Judy looked out the windows, watching *Discovery* grow tinier and tinier as they drove away. She was completely miserable. Not only was she still on solid ground, but she was now wet and freezing. "This is *not* how I thought spaceflight would be," Mike Mullane vented. Judy couldn't have agreed more.

Once they'd dried off, the astronauts learned what the rest of Cape Canaveral had experienced while they were trapped in the crew cabin. The engines' brief ignition had created a bright flash and a large cloud of exhaust that engulfed the Shuttle, followed by the delayed sound of the short engine blast, as it took the sound waves time to travel from the launch site. They expected the Shuttle to climb into the sky, but when it didn't emerge from the exhaust cloud, the spectators had feared the worst—that there'd been some kind of explosion on the launchpad. The crew's families had a brief moment of terror as they thought they were watching their loved ones die. Some screamed while others nearly fainted. A reporter surreptitiously snapped a photo of Judy's mother, Sarah, with her hands over her face, her eyes closed in distress.

Though they all bemoaned their bad luck, the crew eventually found out things could have been a lot worse. The fire NASA detected on the launchpad wasn't your average blaze, but a hydrogen fire, fueled by the same bad valve that had aborted the launch. Hydrogen fires can burn clear, making them invisible to the naked eye. It's possible that if the crew had evacuated the Shuttle when Judy crawled out of her seat, they would have run headfirst into indistinguishable flames lapping up the vehicle's left side. The choice to stay inside may have saved their lives.

Asked by a reporter after the abort how she felt, Judy put the best face she could on the situation: "I was disappointed. But I was relieved that the safety systems do work. It was unfortunate that we had to check them out. But it built confidence in the whole system."

IN THE DAYS just after the abort, the Zoo Crew waited in limbo. They feared the absolute worst: that their flight would be canceled and they'd have to wait for another flight assignment from George months or even years down the line. But there was nothing they could do. They simply had to wait for George to decide their fate.

A couple weeks later, they got their answer. Their flight was still alive, and Judy sang her hallelujahs along with the rest of the crew that they were the closest ones to space. But once again, their payloads would be changing. Now they were acquiring a third satellite to deploy and some additional experiments, originally slated to occur on another flight a few months down the line, STS-41-F. Unfortunately for that flight, it disappeared, and while Judy and her crewmates felt bad for the poor astronauts suddenly without a mission, they all said silent prayers that they were still on the manifest. And they'd still be flying on *Discovery*.

They waited one more month while *Discovery* got a new engine, and engineers packed in the new payloads they'd be carrying. On August 30, they were back at Cape Canaveral sitting inside the Shuttle's cockpit once again. As they entered the T-minus 9 minute hold, the entire crew groaned as they learned that someone in a private plane was flying too close to the pad, potentially triggering an abort. Judy swore under her breath at the pilot's stupidity. Ultimately, he was shooed out of the launch zone, and the countdown continued.

At T-minus 30 seconds, it felt as though this launch might finally happen—though everyone on the crew knew not to celebrate just yet. At T-minus 6 seconds, the engines ignited, and Judy likely held her breath, hoping that they stayed lit.

And then . . . liftoff. After a handful of failed launch attempts, Judy was finally on her way to space, as *Discovery* lurched into the air. She tried to remember her training as the intense oscillations of the launch consumed the vehicle. With no frame of reference down in the mid-deck, all she could do was take cues from the sounds and sensations she felt while staring at a row of white lockers in front of her. If she dared, she could turn her head to the side to look out the window on the main hatch, watching the sky transition from vibrant blue to violet and then finally pitch black. For the first two minutes, it was as if she was in a never-ending earthquake, until the solid rocket boosters broke away, and the surreal smoothness of flight began. Hank did his best to explain every aspect of the launch to Judy and Charlie as the Shuttle sped ever higher, and she almost felt like she was there in the main cockpit with everyone else.

But the men couldn't help but be vague. During liftoff, Hank pointed to some insulation dropping off the external tank and said "You see that?" to Mike Coats. And as the bright Earth came into view in the main windows once they reached orbit, Steve accidentally let slip his awe. "Holy shit!" he cried.

Judy had had enough.

"Knock that off, or I'm coming up there!" she yelled.

Steve's outburst had occurred just as *Discovery* lurched into its final orbit, allowing Judy to actually unstrap. First, she checked on Charlie, who was suffering from a bout of Space Adaptation Syndrome. She told him to stay in his seat a few moments longer, but then she was off to work on her checklist. Her first job: making sure the toilet was up and running. Mike Mullane came to find her down below, and they tumbled through the mid-deck together, celebrating both their arrival in space and the fact that they weren't nauseous.

One of the first things the crew had to do was open *Discovery*'s payload bay doors. It was a crucial maneuver after launch was over. All the electronics in the Shuttle had generated enormous heat, which needed to be radiated out of the cargo hold so that the entire vehicle didn't fry. Fortunately, the doors opened as instructed, but the crew noticed something curious float out of the bay. It was a Coke can. The engineers who'd built the brand-new *Discovery* had inadvertently left their trash behind, including a beer bottle and a roll of duct tape.

Immediately the STS-41-D crew were off to work, with the deployment of their first satellite. It sailed out of the cargo hold as expected, energizing the crew. The next day, however, the second satellite's deployment wasn't as smooth, though the spacecraft itself was fine. The struggle occurred as the crew tried to capture the deployment of Syncom IV-2 with a massive IMAX camera they'd brought on board to film scenes for a new documentary on the Space Shuttle called *The Dream Is Alive*. Judy floated next to Hank as he mounted the massive camera on his shoulders. With the lens pointed out the window, he captured the moment that Syncom floated out into the void.

He continued to film as the satellite left the vicinity of the Shuttle, and Judy leaned in close. Suddenly, she felt an intense, painful tugging on her scalp. She shrieked at the top of her lungs, and the rest of the crew turned from what they were doing to see what had happened.

Judy quickly realized that a good chunk of her free-flying hair had been snagged by the drive belt of the IMAX camera. For some reason, the belt cover hadn't been placed on the camera, and the hungry belt was now actively trying to consume her locks. The men attempted to help her pull out her hair, but the belt continued to munch as Judy screamed in agony. Finally, after enough hair had entered the camera, the strands jammed the machine and the camera's circuit breaker popped.

At last, the IMAX camera was no longer trying to swallow Judy's head, but her locks were still stuck inside the belt. Someone had to grab a pair of scissors to cut the captive strands free. Judy breathed easy again, but now there was another problem. The camera, likely worth as much as one of the Shuttle's components, was too saturated with hair to function. Its motor gears were so thoroughly jammed that the astronauts were a bit clueless as to how to get it working again. It wasn't the end of the world, but it was still a failure that Mission Control needed to know about.

Just as Hank went to radio down to the flight controllers, Judy grabbed him. Seething, she insisted he not mention a *single* word about her hair jamming the camera or else she'd cause him some major bodily harm. And that went for the rest of the crew, too. She implored them to take this secret to their graves.

It took a minute for everyone to understand. Judy knew that the press would have a field day with this if it got out. No one would focus on the success of their mission; instead, they'd all be talking about her and her hair. And since she was only the second American woman to fly in space, she could still feel an intense level of scrutiny placed on her. It wasn't at the level that Sally received, but it was still there; one reporter had asked her during their preflight press conference if any "special arrangements" had been made to accommodate Judy's presence on the flight. She likely predicted that if the IMAX incident became public, a million opinions would form about whether women's long hair should preclude them from going into space.

With Judy's red-hot gaze burrowing into him, Hank picked up the radio to talk to Mission Control. The CAPCOM on duty relayed that they'd noticed a circuit breaker had tripped and realized it came from the IMAX camera.

Hank explained that they were trying to capture Syncom as it was moving away. "We were doing that process, and it jammed," he said, without elaborating further. And that was the end of it. Mission Control moved on and didn't ask any follow-up questions. Eventually, Mike Coats was able to remove enough of Judy's departed hair to get the camera running again.

Unfortunately for Judy, her hair would become a main character on this flight, even without the world knowing of the IMAX incident. Plenty of video footage from the mission was making its way to Earth, showcasing the astronauts attending to their duties in space. At one point, Judy scribbled a message for her father on a piece of notebook paper that read "HI DAD," and thrust it toward a camera filming her. Marvin Resnik was surely thrilled to see the message, but the news anchors and the rest of the world didn't care about the paper. They were mostly focused on Judy's raven-black hair.

Sally's hair had been cut much shorter than Judy's, so her crop had been less buoyant in weightlessness. But Judy's long curly tresses fanned up and out in orbit, looking like a "great cannon cleaner" as Mullane put it. With each turn of her head, the strands lightly bounced and jostled but always maintained a frizzy black halo that followed Judy throughout the Shuttle.

Judy didn't realize just how much of a splash the images of her hair were making on Earth until one morning when Mission Control decided to wake up the crew by playing "Hair" from the Broadway musical. At first she questioned why the song was being played. Then it dawned on her: the world below was talking about her hair. For the next few minutes, she let out a string of expletives, incensed that people were commenting on the last thing she wanted to talk about.

So she focused on her tasks, and the crew successfully deployed their third satellite. Their main jobs were complete. With all the spacecraft out in orbit, it was time for Judy's primary gig: extending the solar array. Equipped with a pair of aviator sunglasses, she flipped the various switches to begin the deploy, and slowly out of the payload bay the array unfurled. Once the 102-foot-long contraption was fully extended, Judy radioed down to Mission Control with an update.

"Roger, Dick," she said, responding to Dick Richards, the CAPCOM on duty. "It's up and it's big."

All throughout training, Judy had joked about making that call once the solar panel was deployed, knowing how it came across. The men teased her about its connotations, but she did it nonetheless, further cementing her status as one of the Six who could appreciate inappropriate humor.

As the flight wound down, Steve noticed a weird reading on one of the computers one day. He was looking at the temperature of an exit nozzle—an opening on the outside of the Space Shuttle where wastewater was periodically dumped. Whenever tanks on board became full of leftover water from the fuel cells, or urine and sweat, the components were dumped out through this nozzle. The valve was equipped with heaters to make sure the exiting water didn't freeze upon release. But Steve noticed that the temperature of the nozzle was low, far lower than it was supposed to be.

Mission Control noticed the temperature reading too. They

wondered if some ice had formed on the exit point as feared. After doing tests to see what was wrong, the flight controllers finally asked Hank to maneuver the robotic arm to the nozzle. With a camera at the end of the appendage, they'd be able to get a visual on what was causing the reading. Sure enough, as the arm swung by the opening, the camera captured a nice chunky icicle of wastewater sticking out on the Shuttle's port side.

Down on the ground, engineers tried to calculate how big the chunk of ice was, estimating it could weigh between 25 to 30 pounds. Hearing that, everyone grew concerned. Flight controllers were worried the icicle could break off during reentry and damage the vehicle's side. The Zoo Crew tried reorienting the Shuttle to put the ice in the sun, but that took far too long. Hank and Mike Coats also tried firing the Shuttle's thruster, shaking the vehicle to and fro to jiggle the ice free. There was even talk of an unplanned spacewalk to remove the ice, which thrilled Mullane. But ultimately, all those options wouldn't work, and Mullane was devastated to learn there wouldn't be an emergency spacewalk. In the end the simplest solution prevailed. Hank simply used the arm to knock off the ice. Sally came into Mission Control during the procedure to offer her expertise.

That didn't solve their problems, though. Worried that dumping more water out of the Shuttle would create another icicle, Mission Control gave the crew an unpleasant command: *Don't urinate in the toilet anymore.* They couldn't risk the urine tank's getting too full and then having to dump the components out of the faulty nozzle.

As an alternative, Mission Control reminded the crew that there were some old plastic urine bags on board—the same kind that were used during the Apollo missions. They could use those to collect their deposits.

However, there was a wrinkle. The CAPCOM on duty informed the crew that they had about "three man-days" left in the toilet tank. That meant *someone* could use the toilet. "We knew we had about three days left in the mission, and so what he was really telling us was *one* person can use it," Steve said. "And without saying it, what he was really saying is, J.R. can use it."

Judy didn't say a word. Nor did she go anywhere near the toilet. "Well, if you guys aren't going to use it, I'm not going to use it," Judy said, as Steve recalled. Again, she did not want to be singled out among her crewmates.

The result was, of course, a mess. In space, pee doesn't simply collect in a bag. Whenever the urine would hit the bag's interior, it'd bounce off the side and break free of the container, floating about the cabin. The crew would have to chase down the escaped droplets to clean them up. Eventually, they realized that stuffing the bags with socks would help soak up the urine, preventing the liquid from escaping. One by one, all the socks on board were sacrificed for the greater good of a urine-free cockpit. At one point, Judy floated by Mullane with a pair of socks on her feet. He noticed the unused socks and grabbed them, attempting to peel the pair off.

"Help, I'm being socked!" she cried.

It would later turn out that Judy's decision *not* to use the toilet was a wise one. They'd learn after landing that the wastewater tank was completely full, and if anyone had used the toilet, they'd have had to probably dump the tank again in orbit, risking another icicle.

Amid all the bathroom chaos, the crew came together to take a group photo. With the men all crowded together, Judy, who was wearing shorts, floated above them, her legs front and center. She'd later receive plenty of mail about her legs being on such visible display.

After just six days in space, the crew returned to Earth, landing at Edwards Air Force Base. Judy was presented with a bouquet of roses. She made sure to hold on to them. A few days after the flight, the crew participated in the customary postflight press conference, debriefing the world on what they'd experienced. Most of the questions were fairly standard, save for one.

"And one more quick question," a reporter from Huntsville asked over the phone. "For any women in the future who might be going up into space, any advice on hairstyles?"

Judy's response made it clear how she felt about that question. "No advice," she said.

Following the mission, engineers at NASA and the agency's contractors did a full postflight analysis of STS-41-D, intimately reviewing the Space Shuttle, the solid rocket boosters, and what was recovered from the external tank. Engineers at Morton Thiokol, the contractor that manufactured the boosters, noticed something odd when they took them apart. They found a tiny bit of soot behind one of the primary O-rings, the thin rubberlike seals that ensured the hot gases inside the booster didn't break free to the outside.

The soot was a sign that the O-ring had failed at its job. The gases had eroded the ring enough that the hot materials actually breached the seal for a short period of time. Fortunately, a secondary O-ring prevented them from escaping into the open air. Still, this was something new. It was the first time engineers had ever seen something like this on any Shuttle flight. They'd seen erosion of the O-rings during flight before—in fact, other O-rings on STS-41-D had been eroded—but never a complete breach of one of the seals.

They called the anomaly the first instance of "blow-by," refer-ring to the hot gases blowing by the seal. But by the time the next launch was imminent, engineers at Morton Thiokol decided that a blow-by of this type was an "acceptable risk," and that Shuttle flights could continue safely.

CHAPTER 13

A Walk into the Void

Above the Chisos Mountains sprawling across Big Bend National Park in West Texas, Kathy sat in the back seat of NASA's WB-57F reconnaissance aircraft as it climbed higher into the sky. The pilot, Jim Korkowski, kept his eye on the jet's altimeter as they ascended. They'd just passed sixty thousand feet, and they weren't done rising. It was a dizzyingly high altitude, but the plane was made to handle such extremes.

Inside the cockpit, both Kathy and Jim were prepared. They were fully outfitted in the air force's high-altitude pressure suits. To the untrained observer, the gear looked almost like actual space suits. Each ensemble consisted of a bulky dark onesie, with thick gloves and a thick helmet. The combination was designed to apply pressure to the body as the high-altitude air thinned away and made it almost impossible for the human body to function.

The duo eventually reached their target height: 63,300 feet. At that altitude, their pressure suits were a matter of life and death. The surrounding air pressure was so low that their blood could start to boil if their bodies were left unprotected. But with the suits on, it was an uneventful research expedition. Kathy took images with a specialized infrared camera that could produce color photos, and she also scanned the distant terrain in various wavelengths of light. They spent just an hour and a half over Big Bend, and the flight lasted just four hours in total.

While it may have seemed a quick and easy flight, Kathy made history when she reached that final altitude above West Texas on July 1, 1979. In that moment, she flew higher than any woman ever had, setting an unofficial world aviation record.

The assignment to train with the WB-57 had scared her at first, but Kathy wound up loving those high-flying planes. "That was very fun, other than this little bit of vague concern that, 'Hope this doesn't mean I'm falling off the face of the Earth,'" Kathy said. The assignment took her on flights up north to Alaska and down south to Peru. As she'd hoped, she received full qualification to wear the air force's pressure suits, becoming the first woman to do so. Soon, donning a full-body suit designed to keep her alive became second nature to her.

NASA officials had also sought her out to test a new piece of equipment they were developing for future Shuttle astronauts, one that would let people relieve themselves while in space. During the Apollo and Gemini eras, NASA developed a relatively complex apparatus for astronauts to pee in their flight suits. It was, in essence, a flexible rubber cuff that fit around the penis, which then attached to a collection bag. The condom-like cuffs came in "small," "medium," and "large" (though Michael Collins claimed the astronauts gave them their own terms: "extra large," "immense," and "unbelievable"). It was certainly not a foolproof system. Urine often escaped from beneath the sheath.

Cuffs certainly weren't going to work once women entered the astronaut corps. While the Space Shuttle had a fancy new toilet for both men and women to use, the astronauts still needed some outlet for when they were strapped to their seats for hours, awaiting launch or reentry. And if one of the women was to do a spacewalk, she'd need some kind of device during those hours afloat. So, NASA engineers created the Disposable Absorption Containment Trunk (DACT). In its most basic form it was . . . a diaper. It was an easy fix in case astronauts needed to urinate while out of reach of the toilet. It was designed to absorb fecal matter, too, though the women probably opted to wait until they reached orbit for that.

Kathy was the best person to test it out. Often during her high-altitude flights, she'd be trapped in her pressure suit for hours on end, creating the perfect testing conditions to analyze the DACT's durability. It worked like a charm. And although the first male Shuttle fliers stuck to the cuffs, eventually the DACT became standard equipment for everyone.

After accumulating hundreds of hours in these pressure suits, Kathy hoped to leverage her experience into a flight assignment, one that might let her take a walk *outside* the Space Shuttle one day. As luck would have it, she ran into Bruce McCandless II in the JSC gym one afternoon. He was *the* guy to know when it came to spacewalks. NASA officials had put him in charge of developing all the spacewalk procedures and protocols, and at times he seemed to live in the NASA pools. Plus, he was always conscripting one of Kathy's classmates to do simulated runs with him in the tanks. Kathy wanted to be next. Projecting as much confidence as she could, she asked him to consider her for his next training run.

It worked. Bruce invited Kathy to accompany him to Marshall

Space Flight Center in Alabama to take a dive in the tank there. The two would be working on spacewalk techniques that might be used one day to assemble a space station. However, the Space Shuttle suits still weren't ready to use yet. Kathy had to wear Apollo moonwalker Pete Conrad's suit, just like Anna had done during her spacewalk simulations. But while the suit swallowed tiny Anna, it was just slightly too small for Kathy, by about an inch. When she put it on, the suit stabbed her shoulders, while parts of it seemed to dig into her chest and back. She tried to stand up and nearly passed out. It took all her strength to walk over to the pool before she flopped into the tank. In the simulated weightless environment, the pain immediately evaporated. But it was still a crucial lesson in space-suit sizes. The suits have to fit their wearers perfectly if the spacewalk is going to work.

The session may have started off painfully, but once she began tinkering with tools and understanding how to maneuver her arms to shift the rest of her body, she was hooked. She loved spacewalking so much that she'd go on to do dozens more practice dives throughout training.

But it wasn't enough to practice in the pool. She wanted to go orbital.

IN THE WEE-MORNING hours of June 18, 1983, Kathy decided to turn on the TV in her California hotel room and watch Sally's launch live. She was preparing to do a dive at Scripps Research Institute that day, to finish her open-water SCUBA certification. But she figured she might as well watch the flight, even if a part of her wished it were her on board.

Her brother, Grant, had tagged along to Scripps to do the dive with her. He sat down next to her and watched the launch broadcast. When it was over, he turned off the TV. "Okay, so you should have been there," he said, referring to the fact that she should have attended the launch in person. Kathy acknowledged that he was probably right and it would have been fine if she'd gone. Then they went off to dive, and Kathy immersed herself in the nearby ocean depths.

Back in Houston six days later, Kathy sat at the large table in the astronaut conference room, listening to audio from STS-7 as the crew came in for a landing. Everyone around her figured the Shuttle wouldn't be landing in Cape Canaveral as planned, not with all the ominous dark clouds looming over the landing site. Kathy eventually left the room and stepped into the hallway, only to be greeted by P. J. Weitz, a veteran astronaut who'd flown a Skylab mission and, most recently, STS-6. "You and I are going to Florida to entertain the VIPs," he told her. "Meet me at the airfield as soon as you can."

The diversion of Sally and her crew to California permitted them to escape the large crowd of officials and celebrities already gathered at Cape Canaveral, eagerly anticipating the return of America's first woman in space. When NASA realized Sally was going to be a no-show, the agency decided to send Kathy in her place—to placate the crowd. Kathy and P. J. quickly jetted off to Kennedy Space Center in a T-38. The moment Kathy stepped into the facility's auditorium, a crowd of thousands descended on her, all buzzing with excitement over this major historic moment. Kathy gulped, feeling completely unprepared to handle the swell of enthusiasm. Everyone wanted to see Sally, and Kathy didn't quite know how to please her fellow astronaut's adoring fans.

At that moment, a thought gripped her. She was glad that Sally had landed in California. That way, Sally would get a few precious hours alone with her crew, before everyone in America—perhaps even the world—grabbed for a piece of her.

"I was just instantly really happy for her that she had this little hiatus to make her own initial sense of the flight—to enjoy it and bask in the moment," Kathy said.

Then Kathy had another realization.

If this is what you get for going first, she can have it!

———

ROUGHLY A MONTH later, Kathy basked in being off the grid. She'd traded in the city of Houston for the Wind River Range in western Wyoming. Some friends had invited her on a nearly three-week backpacking trip through the rocky wilderness. For all, the hope was that they'd forget the stress that came with their everyday lives. For Kathy, that meant escaping the anxiety she felt over the upcoming crew assignments. With the gorgeous view of the white-capped peaks spread out before her, she finally succeeded in pushing Shuttle manifests from her mind.

One day on the trip, Kathy was relaxing on the family farm of one of her travel companions, when she was told she had a phone call. When she picked up the receiver, Sally's voice was on the other end. Sally was still very much in the midst of her nightmare public appearance tour. But she was taking a moment to break some big news: Kathy was going to be assigned to her first spaceflight. The two of them would be flying together on the mission. And Kathy would get to perform a spacewalk.

While it was the news that Kathy had been yearning to hear,

it also came at the strangest time. She'd succeeded in clearing her mind of NASA politics during her vacation. "At one level, of course, I was delighted to hear [of the assignment]," she'd say later, "but at another level I had just so let go of all that and been so immersed that at the moment it was sort of a flat."

Upon Kathy's return, she received the news from George himself. He inquired how she was doing with spacewalk training. She said it was going well. Then he asked if she wanted to do a *real* spacewalk. Seeing as how it was all Kathy had worked for over the last few years, she happily accepted.

But then, George dampened the mood. "The medics think there's a problem with this," he said, referring to the spacewalk. "You better go talk to them."

Kathy later sought out astronaut Joe Kerwin, a doctor who led JSC's life sciences division. He directed her to the culprits: a group of biomedical researchers—all men—who claimed they had research showing women were more likely to get "the bends," decompression sickness, when exposed to low pressures. One place a person experiences lower pressure? Inside a space suit. Kathy thought that sounded crazy. She was an avid scuba diver and didn't seem to have any more trouble with the bends than her male counterparts.

Still, she asked the researchers for their data. It turned out their fears were linked to a study that had been conducted at the Air Force School of Aerospace Medicine in San Antonio. The study compared how flight surgeons and flight nurses fared when removing their oxygen masks inside a pressure chamber designed to simulate the thin upper atmosphere, where pressures are extremely low. Those who wound up getting symptoms of the bends were mostly women.

Based on this study, the NASA researchers had some

recommendations. They felt Kathy could do the spacewalk, but they wanted her to spend more time than her male counterparts breathing in pure oxygen before donning the suit—to completely flush the nitrogen out of her body. All astronauts must do these standard "pre-breathe" periods to lower the chance of getting the bends, but the researchers thought women would need more time to do this.

That seemed pretty ridiculous. What if there was an emergency procedure that required Kathy to get in her suit as quickly as possible? The whole thing was driving her crazy, but she knew the right thing to do was take a hard look at the data. "I managed to keep my cool and not get drawn into feeling I was being accused personally of being unfit for spacewalking or that I was fighting a cosmic battle on behalf of all women," Kathy said. On closer inspection, she found some big flaws. The air force researchers only studied about fifty or sixty people—not a big sample size when it comes to statistics. Also, they didn't take into account other factors that could lead someone to get the bends—body weight, for instance.

After pointing out these issues, Kathy got approval for the spacewalk without any modifications, and finally in November, the mission was announced to the world. Kathy's flight, STS-41-G, would be led by Bob Crippen, the same handsome commander who'd been in charge of Sally's flight. Bob was actually going to be juggling *two* missions at the same time, flying them almost back-to-back. NASA had wanted to see just how quickly astronauts could transition from one flight to another—no small feat when training usually takes up to a year or more. Bob was already in training as commander on STS-41-C, which was going to fly ahead of 41-G. That meant Bob would be absent for most of 41-G's training; he'd only really show up to train about six months before launch.

When George approached Bob about the idea, Bob said he'd only do it under one condition. He wanted Sally on 41-G. "I ended up telling George that if I could have a crew member that I'd flown with before be on that mission and [who] could work with the crew while I was working on 41-C, I thought that we could make it work," Bob said. "And so I asked him to put Sally Ride on the mission." As the flight engineer, Sally would serve as Bob's stand-in while he was gone, almost like a de facto commander. Meanwhile, Jon McBride, a tall, sandy-haired TFNG pilot from West Virginia and a spaceflight rookie, was assigned as pilot.

Kathy's spacewalk partner was Dave Leestma, an astronaut with boyish charm and wavy light brown hair who'd come on board during the 1980 astronaut selection. The pair knew each other well already. They'd both been working on an experiment to see if satellites could be refueled while in space, somewhat like filling up a car's tank at the gas station. There was a higher degree of difficulty, though. Refueling a spacecraft is no small task, especially in orbit where temperatures routinely oscillate from blistering cold to scorching heat depending on the vehicle's proximity and exposure to the sun. On top of that, most satellites run on a nasty type of fuel called hydrazine. The putrid-smelling liquid is wildly toxic to humans. It also has a bad habit of spontaneously exploding if it's heated enough and encounters the right kind of spark. But its combustible nature also makes it a great way to propel spacecraft around in the void.

Because Kathy and Dave had tested refueling approaches while on the ground, NASA now wanted them to see if the procedure could be done in space during a spacewalk. While wearing space suits, they'd connect tanks together in the Shuttle's cargo bay so that hydrazine could transfer between them. The doors of the bay would be wide open, exposing them and the tanks to space. The tanks and mazelike valves they'd be working with would mimic those of a satellite called Landsat 4, an Earth-imaging spacecraft already in orbit that NASA hoped to refuel in the years ahead. In its most basic form it was a tech demonstration, but if it worked, it could open up the possibility of extending the life of satellites by giving them a top-off when they ran low on gas.

In a move that didn't quite make sense to Kathy at the time, Bob made Dave the spacewalk lead, despite Kathy surpassing him in experience and leadership. "I'm a class senior to Dave. I've been in the program longer than Dave. I've worked in the suits more than Dave. I worked this payload longer than Dave did, and I'm number two to him on the spacewalk," Kathy said. "That's really bad optics." But Bob made the decision because Kathy was going to be the lead on another experiment. STS-41-G would be carrying a special imaging radar that would capture pictures of Earth from various angles in orbit. As the resident geologist, Kathy seemed perfect to oversee that project, which in Bob's mind meant Dave had to oversee the spacewalk. But Kathy told Bob that if anyone asked about the assignment discrepancy, he'd have to answer for it.

Regardless of who was lead, Kathy and Dave were going to be spending a lot of time together in JSC's Weightless Environment Training Facility (WETF; pronounced "wet-eff"), which housed a giant tank for conducting dives. Not long after their assignments, the two met in one of NASA's giant hangars to suit up for their first spacewalk dress rehearsal. But once they arrived, they looked around and soon realized there weren't separate rooms to change. Said Dave, "I look across at Kathy and she looks across and sees me and she goes, 'Whoops, is this the wrong place?' And I go, 'This is the only one.'" The area was just one massive open room. And both astronauts needed to fully undress so that they could shimmy into their liquid cooling garments: bodysuits threaded with tubes that would run water all over their bodies. Such garments are critical for spacewalks since they cool down astronauts and prevent them from overheating when stuffed inside their cramped space suits.

The two stood next to each other, holding their bodysuits. They both silently realized what they had to do. In the past, the men would have just stripped down immediately, no questions asked. But for these two, there was a short pause. Then Kathy turned to Dave.

"Dave, let me tell you how I feel about modesty at a moment like this. I have none."

"Fine," Dave said, a bit relieved.

And the two began to slip off their clothing in front of a room full of technicians who'd arrived to help with the simulation. The bystanders, growing pink, quickly dispersed, running for the door to provide at least some privacy. Kathy chuckled to herself as she watched them go.

That first test run was just the beginning of a long and exhaustive training process. Kathy and Dave spent hours upon hours in the pool together, intricately rehearsing the steps of their spacewalk until the movements were practically imprinted on their brains. If anything, they over-rehearsed; the spacewalk itself was only meant to last a few hours. But they wanted to make it perfect.

In fact, a lot of people, including their commander, Bob, didn't approve of the spacewalk as designed. Mostly, he didn't approve of Kathy and Dave using actual hydrazine for the experiment. He would have preferred they use seawater or some other fluid that didn't have a propensity to spontaneously combust. But the space-walkers stood firm. "[Kathy and I] were convinced that the thing to do was use hydrazine," said Dave. "Otherwise, why do it? Because it doesn't really prove anything. We know we can operate it with water." But Crip wanted them to see what they were up against. Just to show Dave how dangerous the fluid could be, he sent him down to the White Sands Test Facility in New Mexico—to witness firsthand experimental detonations with hydrazine.

There in the desert, Dave saw plenty of hydrazine equipment bursting apart into thousands of pieces, the result of researchers not being able to control the fluid properly. The spectacle certainly gave him pause. He and Kathy would have to pressurize tanks with hydrazine inside. The temperature inside the tanks would rise and could spread unevenly throughout the enclosed space. Meanwhile, there'd be no way of knowing where the hydrazine was; in space, it'd be floating all the way through the tank. The concern was that all the rising heat from pressurization might concentrate on a single point, possibly where a whole glob of hydrazine had piled up. If that happened, the tanks would explode.

But Dave left White Sands still convinced the spacewalk could be done safely. He and Kathy came up with a plan. They'd pressurize the tanks slowly for a few moments, then stop, waiting for the temperature to even out inside. They'd follow up with another slow bout of pressurization, again waiting for the temperature to even out. That way, the temperature never reached the point of detonation. They also agreed that they'd have to be as safe as possible. NASA didn't want Dave or Kathy to inadvertently bring any hydrazine inside the airlock with them when they were finished with the experiment. They'd have to stay in their space suits for a few extra minutes while taking readings of the environment with special tubes, making sure they wouldn't contaminate the Shuttle's atmosphere. With all these precautions and more in place, Crip begrudgingly agreed to use hydrazine.

When the 41-G flight assignments were first made, all signs pointed to Kathy making spaceflight history. She'd be the first woman to leave the confines of her spaceship and walk in space. It was a title that even Sally didn't have, though Sally would be making history again too. She'd be the first woman to fly to space for a second time. But when the Soviets heard about these plans, they weren't too keen on the Americans snagging more spaceflight firsts. In July 1984, just months before the scheduled launch of 41-G, Svetlana Savitskaya launched for a second time on board a Soyuz rocket and docked with the Soviet's small Salyut 7 space station. It was a short trip, lasting less than two weeks. But while she was on board, Svetlana donned a space suit and exited the space station. For three hours and thirty-five minutes, she cut and welded metals in orbit with her male crewmate. In one swift weightless metallurgy session, Svetlana denied Sally and Kathy the historic titles they'd been in line to claim.

Back on Earth, NASA was fuming. And managers worried how Kathy would take the news. But Kathy suspected this would happen all along. She'd taken note of the Soviets and their patterns. Whenever the Americans announced something new, the Soviets mysteriously figured out a way to do it first. Once, when a jubilant NASA employee congratulated Kathy and Sally on the titles they'd soon earn, Kathy made sure to lower the individual's expectations.

"No, you have not been paying attention," she said. "It's a very long time between today and that flight date. Let me just promise you a Soviet woman will fly a second flight and get a spacewalk."

Kathy's prophecy easily came true. After Svetlana's walk, Kathy would instead only be the first *American* woman to do a spacewalk. It was a perfectly admirable title, but NASA still tried to figure out a way for Kathy to come out on top. All the spacewalk trainers concocted the idea that if Kathy just stayed out in space slightly *longer* than Svetlana—say three hours and forty minutes—she'd set the duration record. When they told her this, Kathy couldn't help but be amused. She and Dave had trained to hit their timeline as tightly as possible. It felt wrong to purposefully go slower just to make a point. "It didn't strike me as a really big thing to beat her by a few minutes," Kathy said. "I'm certainly not going to go tromping around on dinner speeches or something saying, 'Well yes, but I have the duration record,' because it's all of the same order. It's just silly."

As the launch grew closer, the notion of titles and historic firsts quickly fell away. There was simply too much work to do. During training, Sally had assumed a leadership role for most of the year, doling out instructions to the crew and providing guidance while Bob focused on his other mission. But roughly six months before launch, Bob completed STS-41-C, and he went straight into training with the 41-G crew as they started to get more and more simulator time.

Though the press attention had dramatically calmed down since Sally's flight, plenty of eyes were still on STS-41-G. A lot of the media focus was on the fact that Sally was flying again. They even had a reporter from the *New Yorker* follow them around for their training, documenting their simulation time. Kathy leaned into it. She had a name tag made for herself that read SALLY with a line next to the word. It was a cheeky nod to a Boolean algebra symbol, which meant her name tag technically read "not Sally." Sally didn't find the joke particularly amusing.

But there was still considerable attention on Kathy, given what she was about to accomplish. During the press conferences, Sally, Bob, and Kathy got all the attention. "I mean, there were no questions for Jon and I," Dave said. "So we would just sit down at the end and instead just chuckle."

Just as Kathy expected, one of the reporter's questions was for her and Sally: How did they feel about the fact that Kathy, the first American woman to do a spacewalk, would be in an "observer role, where she'll be observing a man rather than participating directly?" It was a clear reference to Dave's job as the demonstration's lead.

Sally immediately piped in. "I got it," she said to Kathy.

"All yours," Kathy replied.

"First of all, I don't think that's really an appropriate way to put it," Sally began, and she launched into an explanation about how Kathy and Dave were essentially equal partners in the spacewalk. Crip chimed in, too, when Sally was done, but Kathy felt a little vindicated about her feelings toward the uneven assignment.

———————

BRIGHT AND EARLY on launch morning, Kathy and Sally found themselves alone together on the gantry leading to the cockpit, waiting to climb inside the space shuttle *Challenger*. Because of their positions for launch, the two were the last to climb inside the vehicle, giving them a few brief moments to enjoy each other's company before strapping in. They chatted for a bit, but then decided to put on a little show for the cameras by pretending to sync their watches. As they fake fiddled with their wristbands, they joked about what the news commentators might be saying about this exact moment.

Seated next to Sally in the cockpit, directly behind the commander and the pilot, Kathy found herself in orbit just eight and a half minutes after takeoff, relieved that no glitches had gotten in the way. Once in orbit, Crip radioed down to Mission Control to let them know they'd made it without issue. To his surprise, a man with a thick British accent replied. It was a member of the Royal Air Force, admonishing them for using their specific radio frequency. STS-41-G had taken a different path around Earth, a high-inclination orbit that brought them closer to England than previous launches. In that moment, they learned that the RAF used the same radio frequency that the Shuttle always broadcasted on. But before they could come up with a snappy retort, they were already over Africa and out of range.

Unlike Sally's first flight, STS-41-G seemed to be rife with issues. Though nothing was a showstopper for the flight, the crew felt that they were constantly solving problems. Unbeknownst to the general public, this *was* the Space Shuttle's thirteenth flight, which for superstitious types might account for the extra gremlins in the hardware.

The problems began with the lone Earth-observing satellite the crew was supposed to deploy on Day 1. Sally and Dave had been tasked with using the robotic arm to snatch the satellite out of the payload bay and then release it into space. But once the satellite was out of the cargo hold, they realized that its solar panels hadn't deployed properly. Mission Control and the astronauts tried to troubleshoot with all sorts of backup commands, but nothing seemed to work. Finally, flight controllers asked the crew to reorient the Shuttle to point the satellite at the sun. The satellite was still attached to the arm, so it was an easy task.

Just as they maneuvered to start baking the satellite, the Shuttle lost signal with the ground—a frequent occurrence back then when there was far less infrastructure in space for communicating with Mission Control. With flight controllers out of earshot, Sally turned to Dave. "Let's ask Crip if we can try it," she said with a mischievous look. He didn't really know what she was talking about at first, until it dawned on him. She wanted to *shake* the satellite with the robotic arm. "Oh, great!" Dave replied.

Crip's response was that of a lenient parent: "Okay, but don't break anything." So, with Mission Control still in the dark, Dave and Sally waved the satellite back and forth, with much more force and momentum than the satellite operators probably would have liked. But just as they acquired a signal with the ground, the solar panels deployed, with Mission Control oblivious to the impromptu wobbling (though it was likely the sun that did the trick).

The satellite was then sent out into space, paving the way for Kathy to work on her lead project: deploying the Earth-imaging radar. Though the spacewalk was at the top of Kathy's mind, the radar was considered the major reason for performing this mission in the first place. To activate the project, Kathy pressed several buttons and the radar unfolded outward like a blossoming flower. When the first "leaf" of the antenna extended, Kathy's heart rate spiked. The panel began flapping, like the wing of a panicked bird. Such movement was not supposed to happen. And as the flapping continued, the entire antenna began to swivel back and forth.

Unsure exactly what to do, Kathy quickly moved on with the

next phase of the deployment, commanding the other panel to unfold. That seemed to be the answer. The antenna calmed down and remained stable, and Kathy avoided a heart attack.

Unfortunately, the crew's problems with antennas were far from over. On their first flight day, another antenna—this one used for sending large amounts of data to Earth—started to swivel uncontrollably. Mission Control wasn't sure how to fix it other than to cut off power to the offending gimbal system, used to pivot the antenna. That job went to Sally. All it took was removing a cable . . . buried behind a maze of wires and a wall of storage lockers. Sally successfully retrieved the cable, but only after making a bit of a mess in the mid-deck.

The antenna was then stuck in one position, causing a lot of major headaches. To get data down to the ground normally, the antenna would swivel to find a relay satellite that could bounce information to Earth. But since it couldn't point on its own anymore, that meant Crippen had to periodically maneuver the Shuttle to manually point the antenna at a nearby relay satellite. It was a tedious task, but it was the only way for the imaging radar to get the immense amount of data it was recording down to Earth. Meanwhile, Kathy and Dave were constantly swapping out tapes on the high-speed recorders, which would fill up quickly with the radar's data. "We didn't have enough tapes to record all the time," Dave recalled. "So we had to tape it when we couldn't get communications to that antenna and then download it to the ground when we could point the antenna at a satellite."

All of this severely cut into the amount of time the radar could make observations. Adding even more stress to the situation, one of the relay satellites went down for a whole day, possibly zapped by an errant highly charged particle called a cosmic ray coming from deep space. As a result, Mission Control decided to reschedule the spacewalk. Originally planned for Day 5, it was moved to Day 8, to give the radar more time to collect data and beam it back to Earth.

But the move unnerved Kathy. Day 8 was the day before re-entry, and NASA typically didn't schedule spacewalks before the return home. Usually, the day before reentry is a calmer one on the schedule, to help prepare the astronauts for the intense ride back to Earth. And spacewalks are anything but calm.

There weren't any indications that the spacewalk would be can-celed, but Kathy silently worried that it wouldn't happen. She couldn't even begin to think of the despair she'd feel, after the years of hard work she'd put in to get to this point.

But the eighth day arrived, and when she and Dave started getting their space suits ready, she began to believe it was actually happening. Getting inside a space suit is an arduous process. Both Kathy and Dave got a much-needed hand from Jon McBride to suit up. They also did the standard pre-breathe, where they breathed in pure oxygen to reduce their chances of getting the bends. Kathy and Dave took the same amount of time to pre-breathe, despite what NASA's fretting researchers had originally wanted.

When Kathy heard that they were a go for the spacewalk, they were the "sweetest words" ever. "Although you've choreographed all of this, you just feel the momentousness that this is now actually for real," Kathy would later say. Fully encapsulated in her human-shaped spaceship—the same outfit she'd worn hundreds of times on the ground—she followed Dave out the airlock and into the vast open payload bay of *Challenger*. But this time, the full weight of the suit that dragged her down on Earth felt as light as a feather.

"That is really great," Kathy said moments after she stepped outside.

The two floating astronauts barreled forward to the experiment at the back end of the payload bay and immediately got to work. With their tools in hand, they began connecting the two fuel tanks, while taking pictures of their progress.

"I don't need to tell you how much fun this is, do I?" Kathy quipped.

"Not at all," Crip responded.

In the midst of the experiment, they were temporarily stopped by Crip, who radioed the pair to take a quick breather. "Why don't you guys take a break when you finish that picture and take a look at the ground," he said. "We're right over a beautiful part of Canada."

For just a few moments, Kathy gazed out at the magnificent, glowing Earth over her. The sight was beyond words. Only her helmet separated her gaze from the arresting view.

"I'm sorry, make that Cape Cod!" Crip said.

"Oh, look at that," Dave piped in. "Cape Cod is beautiful . . . Long Island, New York."

"There's a lot of Sullivans down there," Kathy replied, referring to her relatives on the East Coast.

It was just a brief glimpse of awe, and then she dove back into the refueling experiment. She had a job to do, after all. For the next three hours, she and Dave did the dance they'd rehearsed to no end on Earth, hitting each step within seconds of what they'd planned. It worked like a dream, a floating ballet against the backdrop of a slowly spinning planet.

With the refueling experiment complete, Kathy had one last task to do before she could go back inside. She needed to address the faulty antenna that had plagued the crew since Day 1. Because it couldn't swivel anymore, she needed to stow it manually in advance of the Shuttle's diving through Earth's atmosphere the next day. So without any tools or restraints—but still tethered to keep from drifting off—Kathy grabbed hold of the edge of the Shuttle, and walked her hands to the antenna's position at the opposite side of the payload bay.

Her journey had to be slow and meticulous since her gloved hands were her only tools to help her "walk" from one end to the next. With her feet pointed up and away from the payload bay, she grasped one part of the ledge after the other, looking perhaps like a circus performer, doing a stunt for an adoring crowd. In the middle of her hand crawl, she heard the voice of Jon McBride in her ear, asking her to hang on a moment. For this flight, the crew had brought the IMAX camera on board again, and he was trying to get the device in the right position to capture Kathy crawling across the Shuttle. Obeying the request, Kathy remained still, looking "down" at her hands clasped on the Shuttle's side. She felt as though she was in a handstand. That was until she lowered her gaze ever so slightly, looking at the Earth below her to see Venezuela and the Caribbean Sea passing by.

In that moment, her perspective suddenly shifted. Recalling it later as a vivid present-tense memory she'd say, "As soon as I move my eyes off my hands and look level and then down a little bit, I feel like I'm not doing a handstand any longer. Now I feel like I'm hanging from a tree limb."

The moment passed once Jon got the camera working, and Kathy proceeded to the antenna to stow it safely for landing. After she and Dave performed a few more antics in their space suits, it was time to come inside. Their recorded time outside was three hours and twenty-nine minutes. It was just shy of Svetlana's time of three hours and thirty-five minutes. A reporter would ask her later if she was disappointed she didn't beat the record. Kathy responded that she didn't care about records. "I could have been the 50,000th or 100,000th woman or human being to do a spacewalk," Kathy said. "In terms of the broad historical backdrop, it still would have been my first spacewalk."

Before they returned home, Kathy spent her spare moments simply watching the view out the windows in the cockpit, staring at the boundary between night and day on Earth—a line known as the terminator that separates the brilliant, illuminated ground from the inky blue and black darkness. On the Shuttle, the astronauts had the distinct pleasure of watching more than a dozen sunrises and sunsets every day, since it took them only ninety minutes to orbit the Earth. The terminator was a frequent sight. But as she looked down at the nighttime side, watching the twinkling lights of the world's cities and infrastructure below, a thought struck her: "Right down there in one of those little patches of light right now, there could be a little girl looking up at the sky and pointing upward and saying to her mother, 'Look, mommy, it's a satellite.' And she's pointing at me."

It reminded her of when she was a little girl, pointing out satellites in the sky to her parents. Now *she* was that satellite in the sky. And perhaps that girl down on Earth could one day travel to space like Kathy did—or even farther.

CHAPTER 14

Anna to the Rescue

B reathing in the chilly air of New York City, Anna stuck her hand out to wave down the nearest taxi. It was February 1984, and she'd just stepped off her flight from Houston at the airport. She was in town, as she was about to go on *The Today Show* the next day to talk about the Space Shuttle mission that was currently in orbit above the Earth.

The mission flying above Anna's head as she stepped into a cab was supposed to be a groundbreaking one. In the next couple of days, one of the astronauts on board, Bruce McCandless, would take the first untethered spacewalk. He'd slip inside his bulky space suit and venture out into the void, without anything connecting him to the Space Shuttle. Instead, he'd be equipped with a new toy—the Manned Maneuvering Unit (MMU)—essentially a jetpack that could propel him through space. It would be his lifeline for getting back to the Shuttle safely. To mere mortals on the ground the concept was unbelievably scary: a lone man adrift in a vacuum. So *The Today Show* asked Anna to appear live on television to talk about Bruce's spacewalk, in case the unthinkable happened and he got lost in space. But if it all went to plan, NASA would produce the breathtaking sight of a suited astronaut floating alone against the vast abyss of space.

Unfortunately, the flight had *not* exactly gone to plan. Before Bruce even donned his space suit and jetpack, the STS-41-B crew had been tasked with deploying a couple of satellites into orbit. Just before Anna hopped on her flight to New York, the crew had released the first satellite, Westar 6—and it promptly failed. It was no fault of the astronauts on board. The satellite was supposed to boost itself to a higher orbit once it was free of the Shuttle, and to do that, it had a helper attached known as a Payload Assist Module (PAM), built by McDonnell Douglas. Equipped with its own engine, the PAM was supposed to fire for a minute and a half, pushing the satellite up to its intended orbit. But that burn petered out after just fifteen to twenty seconds, and the satellite was effectively stranded in space, unable to perform at the lower altitude.

Anna assumed that NASA would hold off on launching the second satellite because of what happened with the first—in case the same failure occurred. Still, she was yearning for an update on the mission as she cabbed to the city. She tested the taxi driver's space knowledge, asking if he'd heard anything about the flight.

It turned out he had. "Yes, they launched the second one and same thing happened," he said.

Anna started to get anxious. Here she was, on her way to *The Today Show*, and NASA had just deployed two satellites that immediately failed. How was she going to explain what happened?

As expected, the bungled deployments were all the anchors wanted to talk about. In the studio, the white-hot lights blazing on her, Anna fielded question after question about the failed satellites.

She answered as best she could, though she was privately baffled about the decision to launch the second satellite. Then, at one point, an anchor asked an interesting question.

Did she think NASA would try to retrieve those satellites?

Anna's knee-jerk response was clear: "No way." She knew the kinds of satellites that the crew of STS-41-B had just released. They were both massive cylinders, each the size of a bus and covered in solar arrays. You couldn't just pluck them out of space and bring them back home, as if you were fetching apples from an orchard. No, she thought, those satellites were staying right where they'd failed.

Anna's prediction would turn out to be wrong. And the failure of those two satellites would be the best thing that could happen for her spaceflight career.

BY THE TIME she'd gone on *The Today Show*, Anna already knew she was going to space. She'd been assigned to her flight in July 1983. By that point, she'd just finished up her duties as lead Cape Crusader for Sally's flight, which had been an incredible experience for her. Anna and Sally got to test out some of the payloads together at Cape Canaveral just before the mission took off. Anna had also been eight months pregnant for that flight. Photographers snapped a few pictures of her working alongside Sally, and a large baby bump could clearly be seen poking from beneath Anna's flight suit.

Back in Houston and about two weeks out from her due date, Anna got a call that George Abbey wanted to see her. And he wanted to see her husband, Bill, too. That struck her as a bit strange, but one didn't question George and his requests. She and her husband slowly trekked out to Building 1 in the summer heat and filed into his office. Anna lowered herself into a seat while George launched into the reason for his summons.

He wanted to assign Anna to a flight. And since she was about to give birth, he wanted to know if she or Bill had any reservations about it.

Anna simply beamed. Going to space had been all she'd ever dreamed of, ever since that day in the schoolyard listening to Alan Shepard's launch at her junior high school. *Of course*, she didn't have any reservations, she told George. And the fact that her boss had enough confidence in her to begin her training with a newborn on the way meant the world to her. She accepted without hesitation. "Oh, definitely, just send me in, Coach," she told him.

Anna's flight would eventually be dubbed STS-51-A, and her crew would be fairly small, including Dave Walker, Joe Allen, and Dale Gardner. The commander would be Rick Hauck, who'd just flown with Sally on STS-7. Their flight would be pretty straightforward, too. They'd be launching a satellite with an Inertial Upper Stage (IUS) attached—the Boeing-built rocket that could boost payloads into higher orbits. To Rick, the flight was "plain vanilla."

During the day, Anna started prepping for her flight plan while waiting for that first sign of labor pains. That entire month, Bill and Anna were on standby for baby Kristin to arrive. Bill maintained part-time work in emergency medicine after work at NASA, but he opted to miss a few night shifts at the ER, just in case the baby decided to come. One Thursday in July, he finally picked up an ER shift after no sign of the baby. That same night, Anna went into labor after a full day of training at JSC. She went to the hospital that evening and after a long labor Kristin arrived Friday morning.

Anna didn't want to stay in the hospital long, though. Having spent a great deal of her life in these sterile environments, she resolved to go home early. Just after Kristin arrived that morning, the nurses moved Anna and Bill to a recovery room. While Bill caught some shut-eye on the floor, Anna tried to get some rest herself—until she looked up and saw her crewmate Dave Walker walking through the door. In his hands, he had a little gift basket with a note: "*Bears for the bairn and the bearer who bore her.*" The gesture touched her heart.

It had only been one weekend since her labor, but Anna felt invincible. So she did something a little bold. On Monday, just three days after giving birth, she showed up at the all-astronauts meeting at JSC, surprising everyone. She brought with her a small doughnut cushion to sit on, but she was there. Her presence made a clear statement to everyone in the room. *I'm here and nothing's going to change.* "I didn't want anyone to think that just because I'd had a baby that I wasn't going to be able to do what I had committed to," Anna said.

Determined to prove her commitment to her job, Anna didn't end up taking a formal maternity leave either. Instead, the training team crafted a unique schedule for her. For the first four weeks, they'd have Anna come in and train roughly two or three days a week, affording her days off to be with Kristin. If she needed to train, she came in. If she wasn't needed, she stayed at home.

It worked well for a time. But then Anna decided to take on another role that would require an immense amount of effort and work. She wanted to be a CAPCOM, the coveted on-the-ground role for any astronaut, relaying commands to crews already in orbit. It was a job that her commander, Rick, didn't want Anna to take. He wanted the crew to hit the ground running with their training, without distraction. But Anna insisted that being a CAPCOM was a vital part of training. She'd know what it felt like to be in Mission Control, and what to expect when communicating with flight controllers on the ground. She told Rick she'd be a better crew member if she did this. He eventually agreed.

So for a time Anna attempted to juggle it all. She trained with her crew, took care of a newborn at home with Bill, and talked back and forth with astronauts in space. On top of all that, she was breastfeeding. And in the early 1980s, NASA management hadn't made accommodations for new mothers—or women in general—much of a priority. Any woman working in Mission Control had to trek to the opposite side of the building to find the restroom. That meant Anna had to sneak away during shifts, journeying to the one place she could discreetly pump milk.

As Anna adjusted to her new roles of both mother and crewmate, she also found herself adjusting to an entirely new mission. One day, the STS-51-A team flew up to Washington for a regular meeting with the Boeing engineers who'd built the IUS rocket booster that would be going up on their flight. When the crew returned to Houston, though, they were informed that the IUS was no longer their payload; they'd be getting a new one. Now they'd launch two communications satellites, Anik D2 and Syncom IV-1. And it was possible they'd be doing something *very* new. Those two satellites that Anna had discussed back on *The Today Show*—the ones she figured were lost in space—could be coming back in a big way.

While NASA had certainly been upset about the failure of the satellites, the insurance companies were the ones that really felt the sting. The two insurers of the satellites, Lloyd's of London and International Technology Underwriters (INTEC) had to fork out $180 million to Indonesia and Western Union, the two operators of the defunct satellites. After making those payouts, the insurers assumed control of the adrift cargo. And they wanted the satellites back. The two entities approached NASA, urging the space agency to form a rescue mission to bring the satellites home. That way the insurers could refurbish the spacecrafts and sell them to recoup some of their losses. Another company, McDonnell Douglas, also expressed interest in the plan. Since the company had built the modules that had failed, it was eager to get them back on the ground to figure out what went wrong.

Bringing back the satellites was a pretty big ask. The insurers wanted NASA to conduct a type of human spaceflight mission that hadn't really been done before. In orbit, these satellites were moving at roughly 17,500 miles per hour, racing around the Earth. A Space Shuttle crew would have to chase them down and match their ludicrous speed with enough precision to retrieve rather than crash into them. Then the astronauts would have to somehow pluck the satellites out of space and tuck them inside the Shuttle's cargo bay for the trip home.

Only one Space Shuttle mission had ever tried something similar. During the STS-41-C mission, astronauts had caught up to a malfunctioning NASA satellite called Solar Max, with the goal of grabbing it and repairing it. The operation was ultimately a success, though the effort to grab Solar Max had been plagued with challenges and during the undertaking a spacewalking crew member was almost lost in space.

Solar Max had an advantage. The satellite had been made to be caught. It was built with a grappling pin, just in case astronauts wanted to grab hold of it one day and fix it while in space. By contrast, the Palapa B2 and Westar 6 satellites were mostly smooth and round. NASA would have to develop something new to latch onto these cylinders—a piece of hardware that had never been made before.

Despite all that, a few things convinced NASA that the mission could be accomplished. The success of the Solar Max job, albeit a patchy one, boosted the agency's confidence. Also, Bruce McCandless *had* successfully demonstrated NASA's new jet backpack, the MMU, showing that the device could safely maneuver suited astronauts through space, no strings attached. For the rescue, NASA pictured an astronaut strapping on the MMU, floating out from the Shuttle, and somehow maneuvering the satellites back to the cargo bay. It'd be like packing up a truck's flatbed before driving home.

It was a risky venture, but to Anna and her crewmates it sounded like a hell of a good time. Getting to try a completely new flight profile like this was the dream. They just hoped the rescue mission would come to *them*. They knew NASA was looking to give the task to an already assembled crew, and they secretly wished it would be theirs. Fortunately, they had a leg up in the competition: their commander, Rick Hauck, had shown off his skills in expertly piloting the Shuttle in proximity to the SPAS-01 satellite—the payload that Sally had grappled with the robotic arm. NASA figured Rick could oversee the same thing again, this time with two big satellites nearby.

Once it became official, the training for STS-51-A went into overdrive. Now they'd be performing one of the most daring spacewalks ever. Dale and Joe would be the two spacewalkers, and they soon immersed themselves in NASA's training pool, rehearsing the intricate orbital ballet over and over. Though Anna would stay inside the Shuttle, she'd have a critical job: using the robotic arm to help snag the satellites and lower them into the cargo bay. The three astronauts would perform as an indoor-outdoor tag team, passing off the satellites to one another as the pilot and commander kept the Space Shuttle steady.

The crew also had to solve one tiny problem. How were they supposed to actually grab these un-grabbable satellites? Dale ultimately came up with an inspired idea. On the back of an envelope, he sketched out the design for a device that became known as "the Stinger." The name was apt, given that it looked like a giant spike. The device would work like a cross between a sword and a plunger. It was designed to be jabbed inside the nozzle of the satellite's engine—the same engine that failed to ignite when the satellites were launched. The spacewalker holding the stinger could then flip a handle and the device would expand, plugging the engine and providing a way to hold on to the massive spacecraft. "It's like opening an umbrella inside a chimney," as Joe once described it.

The team had less than half a year to prepare for the daring salvage, and during that time they had to come up with new tools and procedures that no other crew had used. That meant plenty of training days started and ended without the sun in the sky. Anna practically lived in the simulator that housed the robotic-arm prototype, sometimes bringing Kristin with her for the late-night practice sessions. Once, the toddler sat in a car seat on the floor as Anna worked with training personnel through the night to come up with a new maneuvering routine.

It was an extremely busy time, but the team found room for laughter. Whenever the crew would practice aborts in the simulator, when they arrived at the scenario that had the Shuttle landing in Africa, they'd joke about trading Anna for a pack of Camel cigarettes to get back home. The innuendo didn't bother Anna. "It wasn't anything new to me . . . I understand it's just a way of just breaking the ice," she said. Later, Dave would actually give Anna a collection of various Camel cigarettes in a nod to the joke, another gift from Dave that she cherished.

While Anna never experienced the onslaught of press attention that Sally did, she still became a focus of the mission, since she'd be picking up a historic first: becoming the first mother in space. Apparently, the Soviet Union didn't have among its female cosmonauts a woman they could send up to steal that title.

The press became enamored with this new achievement. "WHEN MOM IS AN ASTRONAUT," one headline declared. It didn't matter that plenty of *fathers* had already been to space before. Everyone wondered how a mom could possibly leave her child for a week to go to space, a telling indication of how the world viewed a mother's priorities. For Anna, it was simple: this was her job. "I had made up my mind what I was going to do, and I never wavered from that decision," she said. When Kristin was first born, Anna received a letter from a "concerned" citizen who felt it important to tell Anna just how irresponsible she was being. She showed the letter to Bill, who tore it up. "Don't read these things," he told her. From that point on, she decided to ignore anything negative that was written about her.

Instead of worrying about the critics, the Fishers leaned into the moment. They had a tiny NASA flight suit made for Kristin, tailored from the same materials that composed the official astronaut flight suits. Anna brought Kristin in her miniature ensemble to NASA for a round of preflight interviews, where photographers snapped mother and daughter together in the simulator's cockpit.

At the end of September, NASA finally told the world that STS-51-A would attempt a rescue of the satellites in space, something that'd been kept under wraps for months. With that announcement, the salvage mission was officially a go. Anna still questioned if the team could actually pull off this stunt. All of them were too nervous to be outwardly confident about their chances. Rick, the commander, saw it as an easy way to fail very publicly. The others simply kept their heads down and continued their grueling training, with fingers crossed every step of the way.

The team began their standard weeklong quarantine on October 31, 1984, though Anna fudged the rules a little on the first day. It was Kristin's first official Halloween, and Anna didn't want to miss such a big milestone. So that first night she left JSC and ran home to go trick-or-treating with Kristin and Bill in their Clear Lake neighborhood. They only went to a few houses, limiting the impact of the indiscretion. Fortunately for Anna, no one at NASA even noticed she was gone.

Anna would run home once more during quarantine—to pick something up. The circumstances were again on her side: The house was empty. Bill, her mother, Elfriede, and the nanny had already left with Kristin to head down to Cape Canaveral ahead of her launch. The sight of the empty house was unexpectedly bracing. Her spaceflight *really* was about to happen—and it was only a few days away.

One last thing Anna made sure to do before her flight was to write a letter. It was to Kristen. The note didn't address the risky endeavor her mother was about to embark on, but the uncertainty of the time ahead. "No matter what happens in the future . . . in spaceflight or in our relationship . . . she came along and gave me incredible joy, a balance and perspective that made me a better person," she said later of the letter's contents. Anna sealed it, with the intention of giving the note to her daughter when she was older—well after Anna had come back from space.

OF COURSE, AT the Kennedy Space Center the quarantine was very much in effect, so even though Anna could see her family from a distance, her heart ached knowing that she couldn't touch her daughter, even though she was just mere feet away. Her mood lifted when one day she went for a routine jog outside the crew quarters to keep in shape. Bill appeared with Kristin in his arms. For a few moments, he matched Anna's stride, so that mother and daughter could be together.

Following that week of separation came the biggest test of all for Anna: the flight itself. She told herself that it was just a week, and that she'd waited for this moment nearly her entire life. But she still hoped to see her daughter one last time before boarding her ride to orbit. Bill tried to make it happen. Since he was an astronaut, he pulled some strings and managed to drive the family onto KSC property in a rental van in the pre-dawn hours before the launch, parking just outside the crew quarters. That way, they'd all get to see Anna up close as she walked out into the open air, just before boarding the Astrovan. Bill had warned Anna's mother that she wasn't allowed to get out of the van under any circumstances. Technically, none of them were supposed to be on-site like this, and there were plenty of photographers around who could blow the family's cover.

Staying in the van was an easy enough request to accommodate at first. But then Anna and her crew emerged from the crew quarters, walking down the ramp and out to the van in front of a gaggle of photographers. Anna moved slowly as she walked, her eyes frantically scanning the crowd of onlookers to see if she could catch a glimpse of Kristin. As hard as she tried, she couldn't find her daughter anywhere. Anna's heart sank. She just had to see Kristin once more before she flew. Just once.

Elfriede looked at her daughter and saw the distress in her eyes. She could tell that Anna was still searching vigorously for Kristin. That's when grandma went rogue. She opened the passenger door and emerged with Kristin hoisted high in her arms. The minor indiscretion paid off. Anna locked eyes on her daughter, a sense of relief flooding through her body. She'd seen Kristin. *Okay, that's behind me*, she thought. Now she could focus on the task at hand.

"Once I saw her, that was fine," she said.

However, Anna wasn't the only one who saw Kristin during the crew walkout. As Bill had feared, a photographer caught sight of Anna's mother with the baby and snapped a picture of them standing outside the van. The two appeared in the pages of a local newspaper the next day. Fortunately for the Fishers, no one at NASA ever realized that Elfriede was standing where she wasn't supposed to be.

But it turned out that Elfriede's rebellious move that morning was all for naught. Anna and her crew climbed into the Shuttle cockpit, strapped into their seats, and then learned that the launch was being scrubbed due to rough winds circulating high above the launch site. The crew had to get out and do their walkout all over again the next day. While she was, of course, disappointed, she worried about seeing Kristin again. *Okay, now what's going to happen tomorrow when we launch?* she thought.

Twenty-four hours later, Anna found herself repeating the

previous day all over again. But this time, she felt like it was actually going to happen. She suited up, ate a light breakfast, and nervously emerged from the crew quarters for a second time. And as she feared, she couldn't find Kristin. This time, Elfriede was prepared. The Fishers had come back in the rental van, and Elfriede got out, holding Kristin up high. Anna locked eyes on her daughter once more while Bill fumed.

Anna was the second to last person to climb inside the Shuttle. As the others took their time getting strapped into their seats, she lingered for a few extra moments on the suspended walkway connecting the launch tower to the cockpit, looking out at the ocean. In those early-morning hours, a full moon shone brilliantly white over the inky-dark Atlantic.

"I cannot tell you what it was like to stand there thinking way back to that time when I was twelve years old, listening to Al Shepard launch, wondering if I would ever have a chance," Anna would later say. "And here I [was] standing and getting ready to launch into space . . . It was such a surreal feeling."

As the flight engineer, Anna sat behind the pilot and the commander, affording her that incredible view of the front windows during flight. Down below in the mid-deck sat Dale Gardner, who was celebrating his birthday that morning, November 8, 1984. Dale made a promise to the flight director over the intercom that he wouldn't "blow out the candle until eight and a half minutes into the flight," when the three main engines would cut out. The countdown ticked down to zero, and as the vibrations took hold of the space shuttle *Discovery* during liftoff, Dave turned back to look at Anna and reached out his hand to grab hers.

Down on the grounds of Cape Canaveral, Kristin watched as

the Shuttle carrying her mother soared into the sky. She pointed up at the glowing dot inching above the horizon. "Oh no, Mama," she said. Once the spacecraft faded from view, Bill asked his daughter "where mama was." She simply pointed up.

JUST AS THE main engines cut off, Anna could feel the blood almost instantly rush to her face. On Earth, the body's liquids pool down in the legs thanks to gravity, but in weightlessness, those fluids shift upward, spreading evenly throughout the head and the extremities. She admired her fuller face in the mirror, noticing that the small wrinkles she spotted at home had miraculously disappeared.

She also felt sick almost immediately. The dreaded Space Adaptation Syndrome had hit her, causing her stomach to lurch. She suppressed both the urge to puke and the temptation to curl up into a ball to take a nap, trying to stick to her assigned duties. Sensing she needed help, Joe came up to her and removed her boots and helmet—giving her one less task to think about. She could still function—about as well as one can with what feels like a sinus infection. She avoided all food for the first couple of days to ensure that she didn't see it again after eating. That night, without a pillow to rest her head, she wondered why she'd pined for space. *I feel terrible*, she thought. *Why did I do this?*

Fortunately for Anna, the first few days of the mission featured the easier assignments. On the second day, the crew deployed their first satellite—for Canada—followed, on the third day, by the second satellite—for the Department of Defense. Both were routine procedures, and the spacecraft lifted out of the cargo bay without any glitches. That third day, Anna experienced her own release. It was as if someone flipped a switch inside, causing her nausea to evaporate. Rick could tell just by looking at her. "Anna is back with us," he said. That morning, she ate a hot dog, and deemed it the best she'd ever eaten—on Earth or in orbit. Now she could start having the space experience she'd always dreamed about.

With the cargo bay emptied, the mood grew tense as the crew prepped for their recovery mission. On the fifth day, Anna and Rick got started just an hour and a half after they woke up. Anna sat next to her commander in the pilot's chair, and together, they calculated all the tiny engine burns Rick would have to do at just the right moments to catch up with the first satellite, Palapa B2. While Joe and Dale donned their space suits, Rick began creeping up on the satellite, advancing like a tiger stalking its prey. With every burn of the Shuttle's thrusters, Rick and Anna triple-checked each other's calculations. At just fifty miles out, Rick was able to view the satellite in his Crew Optical Alignment Sight (COAS), a periscope-like navigational instrument. It looked like nothing more than a tiny little star in the night sky. But soon, that "star" grew steadily bigger and brighter until the full cylinder was in view, slowly rotating in front of them.

Before the flight, Rick had made the call that he was going to get pretty close to these adrift satellites. He parked *Discovery* just thirty-five feet from Palapa, a mere hair's breadth in the expanse of low Earth orbit. The idea was to cut the distance the spacewalkers would have to travel to reach the satellites. With only so much gas in their jetpacks, the astronauts had just seven hours to wrangle each spacecraft and stow it in the cargo bay. Past that point, their oxygen and other critical supplies would get close to depletion. Rick kept *Discovery* steady while the spacewalkers emerged from the airlock.

It was agreed that Joe would be the one to slip free of the Shuttle's confines; Dale would hang back and monitor in the payload bay. As Joe thrust forward inch by inch to greet his target, he held the Stinger in his gloved hands. Without a tether connecting him to the Shuttle, he looked a bit like a white knight participating in the world's most sluggish joust. Inside, Anna and the crew watched through the window, eyes wide, as Joe arrived at the spinning satellite. Its revolutions had been purposefully slowed by controllers on the ground to make the spacecraft easier to handle, but it still turned ever so leisurely. Once at the satellite's back side, Joe stabbed the Stinger inside the engine's nozzle and expanded it, just as the action had been depicted in the drawing on Earth. Palapa B2 was now the world's largest kebob, at the mercy of Joe, who slowed its spin by firing up the thrusters on his jetpack.

With his cargo on a stick, Joe, the smallest male astronaut in the corps at five feet six inches, looked like the world's strongest man grabbing hold of the 1,200-pound Palapa, as he slowly swung the satellite around, putting himself and his prize in just the right position for Anna. At the controls of the robotic arm, she maneuvered the appendage up to the Stinger and wasted no time in latching onto it. "It's interesting to see the large arm coming up right over your shoulder to suddenly grasp the satellite," Joe said. That's when Anna took Joe for a ride. Still holding on to Palapa, he sailed up and over the Shuttle as Anna dragged both satellite and human to the payload bay, where Dale was waiting.

For a while, it seemed like this crazy plan might actually work. But as hopes soared, the team ran into a snag. The spacewalkers had brought with them a special clamp that they were supposed to attach to the other end of Palapa. Once attached, it was to serve as a grapple fixture for the arm so that Anna could latch onto the other end of the satellite and guide the spacecraft down to the cargo bay remotely.

But as Dale tried to attach the clamp, he soon realized the specs were off. The engineers who'd built the clamp hadn't been given the satellite's proper dimensions, and the device was too small by less than an inch. It just wouldn't attach.

It was time to turn to Plan B.

From inside the Shuttle, Dave gave Joe some new instructions, telling him to guide Palapa into the payload bay using his hands— no clamp needed. Always up for a challenge, the diminutive Joe, with his feet anchored to the payload bay, held on to the massive satellite with only his mitts. Meanwhile, Dale removed the Stinger and attached an adapter to the spacecraft so that it would fit snugly inside the payload bay. Joe remained in that pose for a good ninety minutes, arms outstretched, holding the satellite still. Muscle cramps kicked in at one point. But once Dale had attached the adapter, the two spacewalkers used their arms to guide the satellite into the bay, like a key sliding into a lock.

Now knowing just how to ensnare a satellite, the crew did the same procedure on the sixth day with Westar 6, with just a few tweaks. This time, Dale donned the jetpack to joust with and stab the satellite, while Joe, attached to the end of the robotic arm, grabbed hold of the other end to help guide the spacecraft into the bay. The choreography produced the ballet they expected. With both satellites packed tight in the cargo bay, Dale pulled out a piece of paper with "FOR SALE" written in bright red letters. The spacewalking duo posed for a photograph with their bounty neatly tucked in the Shuttle cargo hold.

The mission's success prompted a call from President Reagan, who phoned while the astronauts were still in space to offer his congratulations on a job well done. Addressing Anna directly, Reagan asked if she'd recommend the career of astronaut to her daughter, Kristin.

"Oh, that I would, Mr. President," Anna replied. "The experience is just everything I expected and even more. Seeing the world below us, it makes you realize just how we're all just part of this world. It's a truly incredible experience, and I'm going to recommend it to her highly."

Reporters on the ground also got some minutes with the crew, asking them how the salvage mission went. One European reporter asked Anna if she thought the occupation of astronaut was "compatible with a mother's affection and task."

Anna laughed, answering that she did think it was compatible. "I very much enjoy being a mother; it's a wonderful experience," she replied. "The space program is something that I also believe a great deal in, and I think as we've demonstrated on this flight, there's a tremendous capability there. I don't think the two are incompatible. I think that it's important for me as a person to do the things I believe in, and I think Kristin will benefit from that. Certainly, she'll have a lot of new bedtime stories." She hoped with her answer that she'd exposed the foolishness of the question. No one asked the men, most of whom also had children, if their job was compatible with fatherhood.

With their bounty in the trunk, the crew of STS-51-A returned home, landing at Kennedy Space Center before being whisked away to Houston. Once on the ground in Texas, Kristin was almost immediately in Anna's arms. A wide grin stretched across the astronaut's face as she gripped her toddler in a tight hug.

In the days after the flight, the world's opinions of Anna's flight continued to trickle in. One woman wrote a letter to the editor of the *Philadelphia Inquirer,* commenting that a picture of Anna in space next to a photo of her daughter graphically illustrated "the problems of today's society."

"As I approach my late middle age, I thank God for having a mother who put her family first," the writer opined. "To me, Anna Fisher is not a heroine, but a woman who is derelict in her duty."

Others didn't quite see it that way. Five months later, a New York–based nonprofit called the National Mother's Day Committee informed Anna that she'd be receiving the "Mother of the Year" award along with six other women, including daytime soap opera star Susan Lucci and the governor of Kentucky, Martha Layne Collins. Anna found it ironic that what had earned her this award was leaving her daughter alone for a week. But she wasn't above enjoying the bragging rights that came with the honor.

Later, when Kristin was all grown and understood the tremendous milestone that her mom had achieved, she playfully pointed out that her mother owed everything to her.

CHAPTER 15

The Heist

Sitting in a conference room at JSC, Rhea stared at the television set in front of her, watching as NASA's countdown clock in Cape Canaveral ticked down to T-minus 0. On the screen, the brand-new space shuttle *Discovery* stood erect on the launchpad. She was taking a quick break with some of the other astronauts in late June 1984 to watch Judy's mission, STS-41-D, get off the ground. The room stayed fairly quiet as everyone listened to the countdown. The announcer counted off the numbers one by one, reaching T-minus 6 seconds. *Discovery's* engines ignited as expected, creating an immense cloud of exhaust . . . and then immediately shut off.

Up until that time, everyone in the room had been sitting leisurely around the conference table watching the countdown unfold. Now all the astronauts were on their feet, yelling their opinions at the television screen—Rhea included. Some cried for the astronauts to get the hell out of the cockpit. Others barked back that they should stay right where they were.

No one quite knew the right call. They'd never seen an engine abort like this. But they all breathed a sigh of relief as they watched their fellow astronauts eventually emerge from the cockpit, drenched in water from the fire suppression system. Knowing their colleagues were safe, they all began to wonder what this meant for the Shuttle flight lineup. Rhea was particularly eager to know. Because after this flight, her flight would be next.

Up until T-minus 6 seconds that had been the plan, anyway.

George had beckoned Rhea to his office in August 1983, just a month after Sally had gone to space. When she received George's summons, Rhea wasn't totally sure what it could mean. A small part of her wondered if it might be a mission, but she immediately doubted it. Now a mother of a toddler, she was no longer attending the weekly Friday happy hours, where she assumed most people received their flight assignments from George after having downed a few beers. Instead, her spare time was consumed with chasing Paul through the house while trying to keep up with her medical training. On the weekends, she worked in the emergency room of Sam Houston Memorial, a small hospital in northwest Houston—a way to stay current with her medical expertise. It was a lot of work to add to the mix, and sometimes she wondered if she was really giving NASA the best version of herself. As she trudged over to Building 1, she worried about receiving some type of reprimand.

But George surprised her. After some polite small talk, he got down to it. "Okay. I was wondering if you'd like to fly on STS-41-E next year?"

"Yes, sir!" Rhea accepted, jubilantly.

He told her the flight was scheduled for mid-1984, and she immediately knew what that timing meant. She'd be right in line after Judy. And now with Paul, she'd be making history as the first mother to fly to space—American, Russian, or otherwise. Such titles were just icing on the cake, though. The real gift was that she'd *finally* get to fly, fulfilling the singular goal to which every astronaut aspires. The assignment also felt extra sweet since many of her TFNG classmates hadn't received their calls yet.

Soon she was thrust into a whirlwind of mission training, ER practice, and motherhood. During normal working hours, she was all NASA. Her commander was Karol "Bo" Bobko, who'd flown on STS-6 before Sally. The rest of her crew included Dave Griggs, Don Williams, and Jeff Hoffman. Rhea hadn't been particularly close with any of them before their assignments, but she liked them all well enough and easily got along with them during training. Their flight, STS-41-F (the mission got a name change shortly after), promised to be one exciting ride. They'd be deploying three communications satellites as well as maneuvering an instrument called SPARTAN (Shuttle Pointed Autonomous Research Tool for Astronomy). Like the SPAS satellite that Sally had navigated through space with the robotic arm, SPARTAN was designed to be removed from the payload bay by the arm and released into space, studying distant X-rays. The crew would then retrieve it and place it back in the bay. Rhea was put in charge of the arm for this task, and it was going to be one of the most complicated parts of the mission.

Up to that point all of Rhea's "spare time" had been dedicated to Paul, Hoot, and the ER. Making matters even more stressful was that Hoot was in training himself—for STS-41-B. And adding to the pressure was that Bo insisted Rhea and the rest of the crew have their *own* Monday morning meeting at seven thirty—ahead of the one that started at eight for all the astronauts. For those extra thirty minutes in the morning, Rhea scrambled to find childcare, eventually relying on a neighbor.

She felt like she was carrying the world's largest boulder on her shoulders, and the rock was only growing bigger and weightier with each passing day. And just when she thought the stress couldn't get more overwhelming, George approached her with a new proposition: Would she want to fly a *second* flight in 1986? It seemed like an impossible ask, but Rhea couldn't say no. For this trip, the main payload would be a module called Spacelab, a laboratory designed to fit inside the Shuttle's payload bay and provide a place for astronauts to conduct experiments in space. For one of the few doctor astronauts at NASA, it was a dream assignment.

For months, Rhea tried her best to juggle the two missions, feeling entirely overwhelmed. But then her immense workload suddenly came to a halt. And Judy's flight was the culprit.

As she hollered at the TV that June day, Rhea couldn't have known that that moment would change everything. Following the abort, George and other NASA officials did some quick musical chairs with the Shuttle flights and crews to make sure certain payloads got to space on time. After weeks of uncertainty, Judy's revamped flight absorbed most of the payloads from Rhea's mission. And then one day, Rhea's flight was completely deleted from the schedule. STS-41-F no longer existed.

Rhea couldn't believe it. Her flight had been less than two months away when the abort happened. Now she couldn't say for sure if she'd *ever* see space. The uncertainty was infuriating. "I called my father and ranted," Rhea would later write. "I raged to Hoot until he got tired of listening. I thought about getting in my car and abandoning the whole place." Hoot tried his best to encourage her, but it was hard to find the bright side. "She was pretty miserable. She felt like her whole crew was just the total down-and-out crew," Hoot said. "And I can't blame her."

Eventually, in August, Rhea's torment came to an end. NASA decided that her crew—the one that had been assigned to STS-41-F— would ferry another TDRS satellite that was slated to go up in March 1985. For Rhea, it almost felt like moving to the back of the line. As she trained for the new mission, she watched as, one by one, the other women in her class made the history-making flights she was supposed to make. Kathy became the third American woman in space, with an impressive spacewalk to add to her résumé. And then Anna became the first mother, with her daughter, Kristin, stealing the media show.

"In the grand scheme of life perhaps these weren't major losses, but they were important to me," wrote Rhea. Later, she would express to a friend how the lineup change had disappointed her. Her friend replied, "Whoever remembers any of that?" It was a direct but fair point. Of course, history would remember Sally Ride as the first American woman to fly, but beyond that, which of the Six did what would be quickly forgotten by most.

After getting their new flight assignment, Bo gave the crew a two-week break to relax and regroup. During that time, Rhea reminded herself that she'd still be flying to space, something very few people in the world had ever been able to do. With that fresh perspective, she dove back into training.

———

THEIR FLIGHT MAY have been salvaged at the last minute, but it morphed into something completely unrecognizable. The payloads that they'd trained on for the last year disappeared. Now, they were working with an entirely new type of satellite, one that was so massive it took up all the room in the payload bay. There wouldn't be any room for a robotic arm for Rhea to tinker with.

At the same time, the crew ranks started to swell. A few months into training, NASA added to Rhea's crew a French payload specialist named Patrick Baudry. He'd be bringing a suite of medical experiments with him to space, which certainly interested Rhea. But he wasn't the only one joining.

One day, during a simulation run, Rhea was training for ascent with the rest of the crew on the flight deck when Bo got a call to go see George Abbey. As he left to go talk to the boss, everyone had a feeling what was about to happen next. Bo confirmed their suspicions when he returned. He scribbled a brief note for the crew to read—to keep the news secret from the flight controllers working on the simulation.

Garn has been assigned to our flight.

Garn referred to Republican senator Jake Garn from Utah. The first politician to fly to space had just been put on their flight.

For months, everyone had been wondering which flight Jake might end up on. They all knew he'd be flying at some point. His campaign to fly on the Space Shuttle had begun back in May 1981, just after the first Shuttle took to the skies. Alan Lovelace, NASA's acting administrator, had been testifying before a subcommittee of the Senate Appropriations Committee when, early in the question-and-answer portion, Jake, the committee's chairman, leaned into his microphone and asked, "When do *I* get to go on the Space Shuttle?" He reminded everyone that he had ten thousand hours of flight time as a former pilot in the navy and the Utah Air National Guard.

What he *didn't* have to remind them of was that his committee oversaw NASA's budget. "Any time you are ready to go, Senator," Lovelace replied.

Jake claimed he was mostly joking at the time, but the seed was planted. After that exchange the senator met with various members of NASA management who convinced him that what he wanted was doable. After all, the Shuttle was meant to be this reliable truck that could take anyone into space. Why not a Utah senator? After three years of mulling the decision (and some light lobbying from Jake), in November 1984, James Beggs, NASA's administrator at the time, formally invited Jake to fly on a Space Shuttle mission. He happily accepted.

Most of the astronauts assigned to flights silently hoped the senator wouldn't land with their crew. Having payload specialists fly had already been a tough sell; the astronauts didn't want the responsibility of having to babysit a lawmaker who basically had nothing to do. But when Rhea got the news, she had an optimistic take. "This might be a good thing in the end." At the time, their flight wasn't particularly memorable in any way, and they didn't have a very exciting payload. Perhaps a gimmick like Jake's inclusion would add something sparkly to an otherwise run-of-the-mill mission.

Their flight, however, wasn't done morphing. Just a little over a week before they were set to launch, fate stepped in again. Engineers found a defect in the booster attached to the TDRS satellite they were supposed to fly. They had to take the spacecraft back to the factory to figure out the issue, and their flight entered limbo. Again, Rhea worried that they might see their flight wiped off the manifest. "We felt like the most snake-bit crew that had ever existed," she said. But they were much more fortunate this time around. George told them definitively that their flight was still on. They'd be getting the payloads of a flight set for just a few weeks later. Ironically, they were two communications satellites, the same ones the crew had been originally assigned before the abort that changed everything.

With this new change came another crew shuffle. Four payload specialist positions hung in the balance. Patrick Baudry and Jake Garn were already scheduled on Rhea's flight, but two other specialists, Charlie Walker and Greg Jarvis, had been assigned to the flight they were assimilating. At first, Rhea and the crew thought they'd get Greg, but they wound up with Charlie, whose experiment had already been installed onto *Discovery*, the Shuttle they'd now be flying on. Greg and Patrick were bumped to later flights. They still had Senator Garn, though. He had requested to fly with the original crew he'd trained with, and that kept him on Rhea's flight.

At long last, in April 1985, the crew's flight finally became STS-51-D. "In the end, we would have four different crew pictures, four different crew patches, three different payloads, and two different orbiters," Rhea wrote. It was a long, winding road to reach that point, but they were finally parked at the entrance ramp to space. The robotic arm was back in the payload bay now, with room to store it. And there was another interesting payload that had caught Rhea's eye: an echocardiogram machine, designed to use sound waves to observe the human heart. Medical professionals were eager to look at the heart of a person in space since they knew fluids shifted inside the body in microgravity, possibly causing the heart to change shape. As the doctor on the flight, Rhea took great interest in the experiment and went about recruiting her crewmates to be test subjects.

Quarantine began a week ahead of launch as planned. They were in the home stretch. But after all the intense training, it was the sequestration that was difficult for Rhea. She missed Paul and felt guilty being separated from him. They could talk on the phone on occasion, but with Paul just under three years old, he wasn't a master of telephonic conversation. Rhea just needed to see him—be near him—one more time before she took flight in case the unthinkable occurred.

One day during quarantine, Rhea made an excuse to run to the office. But instead, she hopped in her car and made a quick trip to her home just a couple miles away. There she found Paul in his playroom. She sat with him for just a few minutes before grabbing his face in her hands. Looking into his eyes, she told him she loved him and asked him to remember her face until she came home. It was almost as if Paul could sense her anxiety. He gave her as big a hug as a toddler can muster while planting a big kiss on her cheek.

With tears wetting her face, she left the house to return to quarantine. No one ever knew about her secret trip ahead of the flight.

———

FOR THE MOST part, STS-51-D was turning out to be a pretty boring mission, as Rhea saw it. But in spaceflight boring is often a good thing. The liftoff of the space shuttle *Discovery* was just like any other, with no major glitches or problems. It even took off on time, despite some worrying weather.

Days before takeoff, Rhea had said her tearful goodbyes to Hoot while walking along the shore, a ritual they'd observed before Hoot's first flight too. But now in space, she was focused on her tasks.

Others on board were nursing some serious bouts of Space Adaptation Syndrome, but not Rhea. She didn't feel an ounce of nausea. Jake wasn't so lucky. As soon as he left his seat, he said he felt his "gyros were tumbled"—a reference to gyroscopes, which are used to maintain a vehicle's orientation. Sadly, his sickness had been prophesied before he'd come to space. The senator had asked to be given some sort of useful task while in space, so as to be more than just along for the ride. NASA agreed to make him the subject of medical experiments to see how his body adapted to weightlessness. Some of the experiments involved testing for motion sickness, which prompted cartoonist Garry Trudeau to draw a series of *Doonesbury* comics depicting "Barfin' Jake Garn." Jake had been a good sport about the joke, but now, unfortunately, his life had become the parody.

While he and others tried their best to hold on to their lunches, Rhea was hard at work. On the first day, Dave and Jeff deployed their first communications satellite, Telesat-I, for Canada. Afterward, Rhea worked with the robotic arm to position it to get a good look at the satellite's planned thruster firing. It all went flawlessly, and the crew was feeling proud. Once that deployment was wrapped up, Rhea got her hands on the echocardiogram machine. She used the equipment on herself and then roped in Jeff, Charlie, and Jake as subjects, though it took some persuading to move Jake across the cabin to where the machine was stored.

All in all, things continued to move forward as planned. Rhea tried her best to get some restful sleep—to no avail—so that she'd be ready for the deployment of the second satellite on Day 2. Again, everything seemed to be working fine. Rhea flipped all the necessary switches, pushed all the right buttons to gear up for the deployment. Jeff, the lead on the release, double-checked her work and gave her the thumbs-up. Then she hit the button to set the satellite loose into the void. Through the cockpit window, they all watched the giant cylinder leisurely glide from inside the payload bay and inch out into space.

It was all going exactly as expected. Soon, the Syncom IV-3 satellite would extend its antenna to start communicating with the ground below. The vehicle would then fire up its onboard thrusters, putting it into a spin state to stabilize the satellite and make sure it was oriented properly to do its job. While the frisbee-like satellite increased its spin, it would continue moving farther away from the Shuttle for about forty-five minutes, moving halfway around the world before its boosters fired, putting the vehicle into its higher, final orbit.

At least, that was all *supposed* to happen. But after a little over a minute Jeff noticed something.

"The upper antenna didn't deploy," he pointed out.

Everyone turned and peered out the windows. Jeff was right. The antenna was still in the off position. A few more moments passed but there was no movement. And they all noticed that the satellite wasn't starting to spin faster either. Something was definitely wrong.

Rhea broke the news to Mission Control. "Houston, this is *Discovery*," she radioed. "We are watching the Syncom. The omni antenna is still down."

She knew her words were likely sending the flight controllers into a slight state of panic. A malfunctioning satellite was the last thing anyone wanted to deal with. They asked Rhea to confirm, and she told them again that she and her crewmates couldn't see the antenna, nor could they see the satellite starting to spin up. A few minutes later, after the Syncom had continued to drift farther away, Rhea told Mission Control they were "watching the dark," meaning they couldn't see any fire coming from the satellite. There'd been no ignition. Syncom IV-3 looked dead in the water.

No one was really sure what to do next. But then flight controllers in Houston made an odd request. They told Bo to perform a short burn of the engines. They wanted the crew to stop increasing their distance from the defunct satellite. The Shuttle was going to trail close behind the Syncom.

Uh, what? Rhea and the others were all immediately curious about their new orders. NASA eventually hinted at a possible rendezvous but didn't give any concrete plans for what was to come.

They all went to bed that night feeling a little dejected. Rhea knew that the failure was no fault of theirs. The crew's job was simply to deploy the satellite, and they'd done just that. The stuck antenna was a problem for the satellite's maker, Hughes. Still, it was hard not to feel disappointed. Deploying these two satellites was their mission's primary goal. To have just 50 percent of the cargo working felt like a screwup.

Then the morning came, and NASA gave the crew one heck of a wake-up call.

To save Syncom, Mission Control outlined what sounded like a heist operation. They proposed a plan in which Bo would maneuver the Shuttle close to Syncom, rendezvousing with the satellite like Anna and her crew had on STS-51-A with the Palapa and Westar. Engineers at Hughes speculated that the failure may have been caused by the malfunction of a switch on the satellite's exterior. This little lever was supposed to stay closed when the satellite sat in the bay but pop out when it was deployed. The manufacturers speculated that the switch had gotten stuck. And seeing how it was on the satellite's outside, the astronauts might be able to reach it. Really, it was the only thing on the satellite that the astronauts could do anything about. To activate the switch, two of the astronauts would have to do a completely unplanned spacewalk, fitting some kind of devices on the end of the robotic arm. Then one of the astronauts on board could maneuver the arm with the device attached, using it to flip the switch.

Everyone's mind was racing. It sounded like a completely ludicrous plan. Though it was slightly less ludicrous than what NASA had originally considered. Rather than construct makeshift devices, NASA had thought about simply sending Jeff out in space to flip the switch. "I was going to go out and hold on to the end of the robotic arm with one hand and flick the switch on the satellite with my other," Jeff said years later. "But they decided that was a little bit too sporty." Plus, the commander and pilot would need Jeff's help to navigate during the rendezvous process. So that plan was nixed.

Still, rendezvousing with a satellite was no easy task, and they'd barely trained for the delicate maneuver. They'd done some rendezvous and proximity ops training when they were originally supposed to fly with the SPARTAN payload, but it'd been several months since they'd worked on those skills. As for flipping a switch with the robotic arm, how could they have trained for something like that? No one had ever attempted anything like this in the history of the Space Shuttle—or the entire space program.

But fear and confusion were quickly replaced by charged excitement. Both Jeff and Dave lit up at the thought of doing a spacewalk. They'd trained for an emergency spacewalk on Earth, though they never believed it'd actually happen in orbit. As for Rhea, she'd get one of the most important jobs of all. She'd be operating the robotic arm, using the appendage to snag the stuck switch on the satellite's exterior. Reviving the Syncom would all come down to her skilled hand-eye coordination.

Suddenly this boring launch wasn't so boring after all.

The crew agreed to the heist. Everyone quickly dove into their appointed tasks. First on the to-do list was crafting the makeshift devices that would be attached to the end of the arm. But creating what NASA wanted turned out to be a major headache. NASA flight controllers didn't have the ability to send pictures of what they had in mind to the Shuttle. There was no fax machine on board, just a teleprinter that could print letters on yellow paper. So, people on the ground tried describing what they wanted over the radio. They also sent detailed instructions through the teleprinter, even attempting to illustrate what the device should look like using hundreds of spaced out Xs.

Rhea, Jeff, and Dave got to work on the devices, which quickly turned into space arts and crafts. Using a bone saw from the emergency medical kit, Rhea sawed off part of an aluminum tube used on a circuit breaker, while cutting out holes in the translucent plastic book covers that were used to protect all their checklists. They wrapped the pieces in duct tape while Rhea and Jake stitched some of the parts together with needle and thread, ensuring that everything would stay attached when exposed to the vacuum of space. A sewing kit was kept on board in case there was a tear in a space suit. Rhea recalled her experience stitching up abdomens as she plunged the needle into the duct tape and plastic.

The back-and-forth with Mission Control became a tedious game of telephone, as Rhea tried to shape the materials to NASA's exact specifications.

"We wanted to know if you wanted the thing—the cone—

cone-shaped or if you wanted it folded flat and taped shut?" Rhea radioed down to the ground.

"Go ahead and fold it nearly flat and use the gray tape to draw the sides together," came back Houston's reply.

She'd then take a picture of her progress and send the image to Mission Control. While NASA couldn't send pictures to space, the crew could downlink them to Earth.

Back and forth they went. Rhea and the crew would ask for clarification, NASA would give it, and the astronauts would send down a picture of each incremental change. While it certainly wasn't the most efficient way of working, Rhea couldn't help but have some fun with the project. "It's very interesting," she admitted to Mission Control.

The MacGyvered tools looked like makeshift plastic paddles sporting large rectangular cutout holes. There was one that formed a Y shape, which became the "flyswatter." A second rectangular paddle with wire attached became the "lacrosse stick." Together, the two items would hopefully catch hold of the latch on the outside of the satellite, providing enough force to open the latch. Rhea showed off the finished work to Mission Control, who marveled at how well the crew followed the instructions.

"That's fantastic," the CAPCOM replied.

With their flyswatter in hand, the Shuttle crew became the "SWAT Team," a term coined by Jeff during their crafting session. But of course, even after their successful taping and stitching, their mission was far from over. To help the crew prep for the rendezvous, NASA engineers sent up an entirely new checklist for all the procedures ahead. It was a massive to-do list, and the Shuttle only had the one teleprinter. Long reams of yellow paper containing the instructions spilled out of the printer, filling almost every inch of open space in the cockpit. The resulting sheet of paper was nearly thirty feet long. The crew gingerly collected the paper and cut the rolls into pages. Don taped them onto the pages of another checklist that the crew had already used, making a pristine-looking binder for the upcoming operation.

With their checklist fully printed, they began focusing on the unplanned spacewalk. On Day 5 Jeff and Dave jubilantly wiggled into their space suits and grabbed hold of the flyswatter and lacrosse stick to take with them outside. To attach the devices to the end of the robotic arm, they'd use a special strap that was stored on board, provided in case the crew needed to strap anything down in the payload bay. Out the airlock they went, just as the sun was setting behind the Earth. A deep red hue showed across the open bay.

As Rhea and the others looked out the window, they couldn't believe this was actually happening—a completely unplanned space-walk, the first in Shuttle history. In fact, when Dave and Jeff had been training in the pool, Jeff had joked that he'd throw a beer party for all the trainers if they actually did a spacewalk. As he stepped out into the void, he thought to himself, *We're going to have to pay up.*

The two spacewalkers walked on their hands, carrying them-selves out to the end of the robotic arm, which Rhea had posi-tioned close to the payload bay. Since all of this was completely unplanned, the spacewalkers didn't have any foot restraints or special tools to aid them in their journey. Though they were able to tether themselves to the Shuttle to keep from floating out into space, the duo had to make do with their own appendages for support, steadying themselves with their hands against the Shuttle's edge. In microgravity, such a task isn't as simple as it may seem.

The wrong turn of a wrist can be enough to send a spacewalker twirling in the wrong direction.

It took them about an hour to attach the flyswatter and the lacrosse stick. Afterward, NASA wanted to make sure that the robotic arm would still fit in the payload bay with its new attachments. Otherwise, they'd have to do a second spacewalk to remove the devices. So, Mission Control asked Rhea to see if she could fit the arm inside the bay. But at the time of the request, the sun was setting again and the light wasn't great for visibility. Mission Control asked Dave and Jeff if they could stay outside for another forty minutes to wait for sunrise.

"Well, I'm sure you know how tough that's going to be," Jeff replied, elated.

Rhea watched as they proceeded to crawl all around the payload bay, relishing every minute of extra playtime. They even crawled to the cockpit windows to get their picture taken. "So there I was, out in a space suit, just watching the world go by with nothing to do," Jeff said.

With the sun shining on the orbiter again, Rhea showed that the arm fit just fine with the flyswatter and lacrosse stick attached, and the spacewalkers returned to the cabin with the widest of grins on their faces. All the crew felt it: this space heist might actually work.

At last, the big day arrived: rendezvous day, when a verdict on their efforts would be rendered. Everyone wondered if Bo and Don could actually maneuver the Shuttle close enough to the Syncom satellite without ramming into it; they'd barely trained for such a scenario. The pressure was also on Rhea, whose job it was to use the arm to tug Syncom's switch. If all went well, the lever would tear through the plastic devices she and the others had made, but the force would be enough to pull the switch into the right position. At least, that was the plan.

Bo and Don began to maneuver the Shuttle closer to Syncom. The satellite was so far away at that point that they couldn't even see it in the void. But as the Shuttle periodically fired its thrusters, Syncom soon appeared on the horizon, a tiny spec almost indistinguishable from a star. Bigger and bigger it became as the commander and pilot brought the spaceplane closer to the defunct satellite. Soon Syncom IV-3 loomed large in the field of view, a massive cylinder with its booster pointed straight at the Shuttle. "Somehow this was the moment that I started taking this operation seriously," Rhea would later write. She knew that if that booster ignited, they were toast. The entire crew was banking on this satellite staying calm during their operation.

With Syncom right next to the Shuttle, they waited. The timing of the operation had to be precise. If the switch was flipped at the right time, Syncom would burn its thrusters forty-five minutes later—part of its original plan. And NASA wanted to be sure the satellite was in the right spot in space when that burn took place. Rhea had a window of only a few minutes to snag and flip the switch.

As she waited, she pinpointed the target lever on the satellite. There it was, situated in a bright orange square that made it pop against the darkened satellite. Using the robotic arm's camera, she captured an image of the lever for the monitor in the cockpit. At first, the sight was a little disheartening. The switch looked like it was in the on position. Still, she resolved to make the snag. Maybe an extra tug was all the satellite needed.

The window came for the maneuver, and with a flick of the wrist and a turn of the knob, Rhea slowly put the arm in position—its two makeshift devices outstretched. Syncom IV-3 was still turning, and she had to pull on the lever just as the spacecraft rotated into the right position.

The moment came, Rhea inched the arm forward, and . . . success! The flyswatter caught on the switch, tearing through the plastic. To ensure success, they waited for the satellite to do one full turn and snagged the lever a second time with the lacrosse stick. The switch tore the wire, just as predicted. Syncom came around again for one last try. "We got, as we can count, at least three really good whacks at it," Rhea said after the mission.

Rhea had done it. She'd completed the space heist! She and her crew had executed NASA's instructions flawlessly.

Unfortunately, it wasn't enough to revive the satellite. Whatever had prevented the antenna from deploying and the booster from igniting had nothing to do with the switch. It must have been some internal bug that the engineers couldn't identify, and if it was on the inside, there was nothing the astronauts could do. The Syncom continued to spin slowly out into space, never picking up speed or performing the burn it was supposed to.

Still, Rhea and her crewmates were beaming with pride for what they'd accomplished. They may not have saved the satellite, but they'd completed a complicated mission never before performed. Their reward was an extra day in orbit, which they spent playing with children's toys like slinkies and yo-yos, a fun experiment to show kids how these objects behaved in weightlessness.

Before long, Rhea and her crew were on their way to Cape Canaveral, coming in steep and fast on the concrete runway at Kennedy Space Center. Sitting in the mid-deck, Rhea only had the small window in the main hatch to give her clues as to their location in the sky. The touchdown was smooth, but to compensate for the intense crosswinds, Bo heavily applied the brakes to keep the Shuttle from blowing off course.

That's when Rhea heard a loud bang, which felt like something had burst beneath her seat. But with no warning lights indicating a problem, they pulled to a stop and emerged from the Shuttle. That's when they learned they'd shredded a tire, the first time the Shuttle would blow one during landing.

Hoot was waiting for Rhea outside the Operations and Checkout Building at KSC, and Rhea felt sweet relief wrapping her arms around him. But there was one more man she needed to see. They quickly jetted off to Houston, where little Paul was waiting to see his mother.

With her feet back on Earth, Rhea learned of all the press coverage her flight had received—not simply because of Jake Garn but because of their impromptu salvage effort. Later, Anna told Rhea that a couple of the astronauts—self-proclaimed experts on the robotic arm—doubted Rhea's ability to pull off the rescue stunt. She never said an angry word to them. She felt that her success was already the greatest form of revenge.

But during the flight there had been a person at NASA who'd stuck up for Rhea. As she stitched together the lacrosse stick and flyswatter, someone in Mission Control made a comment over the radio that Rhea was "a good seamstress." Sally Ride, who'd stopped by to see if her expertise might be needed and overheard the remark, tapped the person on the shoulder and corrected them.

"Good *surgeon*," Sally said.

CHAPTER 16

The Prince and the Frog

After years of waiting without one complaint, it was Shannon's turn at last.

Shannon wasn't one to seek the spotlight, nor did she get caught up in the politics of jockeying for an early flight assignment. She came to NASA as a working mother of three, and that meant skipping after-work social events that included Friday night happy hours, where the astronaut corps got precious face time with George. When the other women would get together to discuss various concerns related to their group, she'd usually be absent, needed at home to cook and take care of the kids. "After hours, I couldn't do much," Shannon recalled. "I mean, I was home. I didn't have a whole lot of outside time to do social things." All the while, she kept her head down and worked hard at whatever job she was assigned, hoping it would lead to an assignment.

Finally, her perseverance paid off one Saturday morning in late 1983 when George phoned her at home to see if she'd like to fly. Excited to finally fulfill her childhood dream, Shannon said, "Of course!" But George didn't tell her much else. She had to sit on that information and wait until Monday to learn who her crewmates would be. That morning at JSC, she went into the office she shared with John Creighton, trying to keep quiet about her assignment. But John was one of the people she *most* wanted to tell her news to. The two had been officemates since they first arrived at NASA, and a great friendship had formed between them. They'd often tease each other, especially about their hometowns. "She loved Oklahoma," John said. "She used to tell me that's God's best place to live. And I kept telling her, 'No that's Seattle,'" which John considered his hometown.

John was acting somewhat strange that morning, too, as if he was hiding something. Shannon figured something was up. Finally, she asked politely if he happened to get a flight assignment recently. He said he had, and she spilled the beans about hers. They both realized they were on the same flight. Shannon couldn't have been more thrilled. Not only was she getting to fly, but she'd be sharing the experience with a great friend.

They soon learned who'd be filling out the rest of the crew. Their commander would be Dan Brandenstein, another TFNG who'd already piloted STS-8. John Fabian, who'd flown with Sally, would make his second flight, and pilot Steve Nagel would round out the list, though for this trip he'd serve as mission specialist.

But just as Rhea's flight had suffered from ever-evolving identities, Shannon also saw her flight completely transform in the years following her assignment. Originally labeled STS-51-A (the eventual moniker for Anna's flight), the mission was supposed to launch in October 1984 with a communications satellite and a special research lab on board. That had completely changed by the next year. By August 1984, the flight had become STS-51-D (which eventually became the name of Rhea's flight). Now, they were supposed to launch one big Syncom satellite, as well as do a rendezvous with a bus-sized cylindrical spacecraft called the Long Duration Exposure Facility. The LDEF, which had been placed in orbit by the crew of STS-41-C in April 1984, was now supposed to come back home, and Shannon's team had been tasked with picking it up. They'd also gain some payload specialists along the way: Charlie Walker from McDonnell Douglas and Greg Jarvis from Hughes Aircraft. Shannon enjoyed the additions, though, and she grew closer with Greg in the months they worked together.

But their flight was about to go through some serious upheaval, again. Just a month before they were scheduled to fly, the crew did the obligatory press conference for reporters, discussing all the details of how they'd deploy the Syncom and rendezvous with the LDEF. Once the press conference was over, they walked back to the Astronaut Office, where they learned that their flight had completely changed. They'd no longer be flying with the payloads they'd just discussed with the media.

The instability of Rhea's flight ahead of them had reverberated onto them. When the TDRS satellite was taken off the manifest for Rhea's mission, the STS-51-D crew absorbed the Syncom satellite that Shannon's crew was supposed to launch (which ultimately led to Rhea's space heist). That led the two crews to do a payload specialist swap. Shannon's crew took the French payload specialist Patrick Baudry from Rhea's flight, while Rhea and her crew wound up taking Charlie Walker.

Greg Jarvis wound up getting bumped to another flight down the line. Shannon was pretty upset at losing a friend. But such was the way of the unstable Shuttle manifest at the time. Change was a constant, and the astronauts just had to accept it.

After all the uncertainty, Shannon's crew finally settled on a new mission called STS-51-G, and this one would stick. They went from flying one large satellite, Syncom, to flying three smaller communications satellites. One of those satellites was called Arabsat-1B, a spacecraft primarily funded by Saudi Arabia that would provide communications services for the countries of the Arab League. And with the Arabsat's addition to the manifest, the crew would be getting a *new* payload specialist—one handpicked by Saudi Arabia to "oversee" the satellite's deployment.

It was March 1985, and the crew was just three months from takeoff with an all-new set of payloads and new payload specialists. And they *still* didn't know who the Saudi Arabian flier would be. With little time until launch, Saudi Arabia conducted a hasty search to fill the seat, limiting the selection pool to pilots who could speak fluent English.

It may have seemed a bit fast and loose, but by this point in the program, non-astronauts and non-payload specialists were riding on the Space Shuttle with increasing frequency. NASA had long portrayed the Space Shuttle as a simple truck to orbit—one anyone could ride safely, not just NASA astronauts. And after years of promoting this line, the agency was finally starting to follow through. Jake Garn's successful flight with Rhea would prompt NASA to start making preparations to fly another politician, Democratic representative Bill Nelson of Florida, on an upcoming flight. As the chairman of the Subcommittee on Space Science and Applications, Nelson was essentially Jake's equivalent on the House side, and NASA had expressed interest in flying all the chairs of NASA's appropriations and authorizations subcommittees.

Adding to the number of people looking to pass through the

Shuttle turnstile: Seven months earlier, in August 1984, President Reagan had announced the beginning of a new nationwide search, this time for the first citizen astronaut to ride aboard the Shuttle. Except it wouldn't be just *anyone*. "I'm directing NASA to begin a search in all of our elementary and secondary schools and to choose as the first citizen passenger in the history of our space program one of America's finest—a teacher," Reagan declared. But NASA wasn't going to stop there. Preparations would soon be made for a follow-on program—one that would send the first journalist into space.

Plenty of astronauts thought this "anyone can fly" trend was moving in the wrong direction. Whenever a seat was given to a payload specialist, a politician, or a civilian, that meant a seat was *not* being given to an astronaut, a person who'd spent years training for the sole purpose of going to space. Plus, a number of the astronauts believed NASA was making the mistake of believing its own hype. Not a single astronaut considered the Shuttle to be completely routine and fully operational with little chance of failure. Many held concerns about the safety risks of putting inexperienced riders in a coveted Shuttle seat.

While the idea of a newcomer from Saudi Arabia entering the mix at the last minute might have made some feel slightly uneasy, there were also concerns about the payload this person would be "overseeing." "Arabsat had three safety reviews. It never passed any of them," John Fabian said. "Mr. Abbey didn't think we should fly it. I didn't think we should fly it. The flight controllers didn't think we should fly it. And the JSC Safety Office didn't think we should fly it. And yet, we flew it." At the time, NASA was in an undeclared race with Europe's main launch provider, Arianespace, to launch as many communications satellites as possible. And John argued

329

that boundless confidence had, by that time, become part of the agency's brand. "NASA didn't make mistakes" back then, John said. "We were invincible."

So the crew pushed on, training for their entirely new mission in just three months' time. In addition to what was already on the astronauts' plate, Dan scheduled a lesson on Saudi Arabian culture, so that the crew would know how to interact with whoever was picked to join them in orbit. He asked representatives from Aramco, the Saudi Arabian Oil Company, to come to JSC and give the crew a briefing on Saudi customs.

"We wanted to make them feel welcome and not do anything that would be offensive," Dan later recalled. "None of us knew the culture." The way Steve remembered it, the meeting revolved around some basic things *not* to do: "No camel jokes. No harem jokes. Don't do that around them," he recalled being told. The crew took that advice to heart.

Finally in April, Shannon and her crewmates learned who their seventh crewmate would be. And it turned out to be a royal surprise. The overseer of the Arabsat would be the twenty-eight-year-old Sultan bin Salman Al Saud, a Saudi prince and nephew of King Fahd of Saudi Arabia. His presence would make him the first member of a royal family to fly in space—as well as the youngest person to fly on the Space Shuttle. Many wondered if nepotism had won him the role, but Sultan maintained that he'd actually had to convince his family to let him accept the ride.

While Dan had done his best to prep the crew for interacting with an individual from Saudi Arabia, Sultan turned out to be fairly familiar with America already. He spoke English fluently since he'd lived in the United States for much of the 1970s. He studied mass communications at the University of Denver—chosen because he wanted to be close to skiing. In his spare time, he'd taken up flying privately and received his pilot's license in 1977.

Having had that prohibition against "camel jokes" drilled into their heads, they were all on their best behavior when Sultan first introduced himself in Houston just a couple of months before launch. But his willingness to poke fun at himself confounded their lessons in cultural sensitivity.

"Lo and behold, the first time we met Sultan, the first thing he did was tell a camel joke," John Creighton said. The Saudi remarked that he'd left his camel outside on the way in. It was the perfect icebreaker, and they quickly realized there was no need to walk on eggshells. If anything, Sultan understood American culture better than anyone. Sometimes during training, one of the crew members would make a joke, which the Frenchman, Patrick, wouldn't understand. Sultan would sometimes lean over to fill him in on the subtext. Over time, the TFNGs gave STS-51-G a nickname as a nod to their international diversity: The Frog and the Prince Mission.

Though Sultan may have been at ease with his American crewmates, Shannon was reserved with the prince at first. She wasn't a big fan of Saudi Arabian culture, especially when it came to the treatment of women, so she kept her distance in the first few weeks of training. But over time, she warmed to Sultan and they became friendly and professional, though never completely open with each other. Because Shannon was a woman and Sultan came from a country where women never would have been considered for an endeavor like spaceflight at the time, a culture clash seemed unavoidable. But none of the awkwardness emanated from Sultan himself. Rather, it came from the various NASA managers and Saudi officials desperate to avoid committing a cultural faux pas.

The first hiccup occurred during the last hectic weeks of training and stemmed from an experiment that the French had put on the itinerary, which involved generating echocardiograms similar to those Rhea had produced on her mission. The researcher in charge of the experiment co-opted Shannon into being a subject while on board. Patrick would scan Shannon's heart with the specially designed device that'd he brought along. But the researcher wanted more subjects than just Shannon for Patrick to test the equipment on.

She thought for a moment. "Hey, see if Sultan can do it, because he's not doing anything now. He'd be the perfect subject." In space, Sultan's biggest task would be to take pictures of the Arabsat's deployment, as well as to snap photos of Saudi Arabia and the Middle East. The researcher liked the notion and went off to consult NASA officials about it. The next time Shannon saw him, though, he gave her some bad news.

"Shannon, that was a great idea, but Sultan can't be [a subject]."

"How come?" Shannon asked.

He explained to Shannon that they only had one probe for the experiment, which meant the same probe that touched Shannon's skin would have to touch Sultan's.

"And we can't have a probe that touches your body touching a royal body," he explained.

It felt a bit like Shannon had been thrust back into the early days of her career. The feeling intensified when she was told not to wear shorts during the mission, so there wouldn't be too much of her legs showing when pictures were taken. It was a recommendation Shannon decided to ignore. She was going to do her job and she didn't care if how she did it offended anyone.

―――――――

THREE WEEKS BEFORE launch, the crew lay on their backs inside the space shuttle *Discovery*, staring up at the Florida sky through its windows. They'd been conducting obligatory dress rehearsals, going through the motions of pulling up to the launchpad, climbing into the cockpit, and strapping in, just as they would do on the day of the flight. It was tedious work, with plenty of long hours filled with waiting while strapped to a hard metal seat. For all the flying astronauts got to do, there was just as much waiting to do on the ground.

While they lay horizontally in the cockpit, John Creighton looked over toward his crewmates and noticed Shannon had her eyes closed. She was fast asleep, dreaming the time away. "We started talking quietly," John recalled, "and we kept saying, 'Do you think she's going to do that on launch morning?'"

Sure enough, when it came time to launch for real on June 17, Shannon again lay back in her cockpit seat and took a short nap while waiting for the countdown clock to reach zero. That was just the way Shannon operated. She was cool, calm, and collected—a woman comfortable with her situation. She was perfectly content there on *Discovery*, so much so that she snuck in some shut-eye until it was time for the real work to begin.

But ten minutes before the clock reached zero, Shannon's eyes were back open. She was wide awake as the Space Shuttle's engines ignited, and the vehicle lurched off the launchpad, hurtling into the sky. Of all the things Shannon thought at the time, the two words that stood out were: *at last*. "I mean, it was such a relief to think that you're actually on your way," she said, having dreamed of this moment as a little girl when she saw Sputnik arc across the sky.

Soon it was time to unstrap, but before leaving his seat, John Fabian pulled out a special item he'd brought with him. It was a lock, the first of its kind to ride on the Shuttle. The decision had been made ahead of the flight to include such a device on the Shuttle's outer hatch, due in no small part to the variety of travelers that NASA had been taking to space these days. Recalled John, "There was an incident on a previous flight—and I won't get involved in personalities at all—that led Mr. Abbey and others to decide that we needed to have a safety mechanism." Questions arose about a passenger's emotional state, and NASA didn't want someone like that to be in a position to become "infamous." Going forward, there'd be no opportunity for a frazzled astronaut to swing the door open and expose everyone to the deadly void.

On each of the first three days, Shannon and the other mission specialists deployed the three communications satellites: first, one from Mexico, then the Arabsat, and then a communications satellite called Telstar-303. The operations were a breeze, with each satellite leisurely floating out into space and functioning as expected.

On the fourth day, Shannon put her robotic-arm skills to work. Just as Sally and John Fabian had done with the SPAS-01 payload, Shannon manipulated the Shuttle's famous Canadian appendage to remove a payload from the cargo bay, called the Shuttle Pointed Autonomous Research Tool for Astronomy, or SPARTAN. It was the same payload that Rhea and her crew had trained for, but it got passed on to Shannon. Designed to study X-rays and the super-massive black hole at the center of the Milky Way, SPARTAN was a free-flying payload meant to operate on its own, disconnected from the Shuttle's main interface. With ease, Shannon lifted the more than two-thousand-pound SPARTAN out of the bay with the robotic arm, dropping it out in space to live on its own for the next forty-five hours, staring at our universe's most enigmatic celestial objects. Once the satellite had fulfilled its role, the robotic arm was used to retrieve SPARTAN and tuck it back into the payload bay.

Throughout the payload deployment Shannon remained calm and collected. No one worried about a task when Shannon was at the helm.

"You never have to worry about whether or not something's going to be right, because Shannon is going to make it right," John said, looking back. "And that's not true of everybody. That's not true of some of the finest people I've known in my life, because if they get dressed in a hurry, they might forget something. And Shannon is not going to do that. She's going to do it exactly the way it says."

While Shannon and the others stuck to their work in orbit, everyone on the ground was, of course, interested in Sultan. He'd made his observations of the Arabsat while taking pictures of the countries in the Arab League as the Shuttle passed overhead. Though he'd brought some science experiments on board with him, he spent a large chunk of his free time reading the Quran and praying. Technically, as a Muslim, he was supposed to pray five times a day toward Mecca, but up in orbit, astronauts loop the Earth sixteen times in a twenty-four-hour period—and quickly after Mecca is in sight it will be in the rearview mirror. Rather than pray five times every orbit, he received a blessing to pray three times a day instead based on Florida time. He also became the first person to fast a day of the month of Ramadan in space, with the Islamic holy month coming to an end a day after the launch.

With so many eager to see the royal in space, NASA scheduled the crew for a televised press conference on the last day of their mission, and once again, NASA worried about its image ahead of the event. Shannon had been wearing shorts throughout the orbital voyage, just as she'd planned. Pictures of her had been coming down to Earth, and NASA worried about Shannon's bare legs floating next to Saudi Arabian royalty on television.

An official message was sent to Mission Control in Houston, requesting that the crew wear pants for the press conference. Mike Mullane, who was serving as CAPCOM for STS-51-G, took the note. A former self-proclaimed sexist pig, Mike saw the message and immediately threw it in the trash.

"I wanted to call Shannon and tell her to wear a thong for the press conference," Mike wrote.

Despite NASA's concerns, the press conference was just fine, with Sultan proclaiming that the last few days had changed him. "Looking at it from here, the troubles all over the world and not just the Middle East look very strange as you see the boundaries and border lines disappearing." As for Shannon's legs, they were out of frame the entire time, concealing her attire.

STS-51-G turned out to be a nearly flawless mission without the major trials that some of the previous missions had suffered. The crew returned home on time, and as Shannon put her feet back on Earth again, she felt as if she'd doubled her weight while in space—a common feeling among many who transition from weightlessness to our planet's gravity.

"I thought, 'Oh, my goodness, I feel so heavy,'" she said. "You mean, I have to live the rest of my life just like this?" Eventually, the weighted feeling subsided and Shannon flew back to Houston to see her family. There, her husband and children were waiting for her.

As her youngest child ran up to her, Shannon braced herself for words of love from the boy.

"Mom," he cried, "what are you going to cook for supper tonight? I'm so tired of pizza!"

———

ULTIMATELY, THE FLIGHT to space had been the smoothest part for Shannon. It was *after* the flight that things took a more dramatic turn.

As part of the routine postflight press tour, the crew traveled to France, the home of Patrick, as well as cities throughout the US. But also on the itinerary was a planned trip to Saudi Arabia.

It was trouble from the start.

The planned schedules had all the men attending events together while Shannon engaged in a separate activity. Shannon reminded the trip planners that she was part of the crew and needed to be with everyone whenever they appeared at events. And then there was the issue of entering the country. At the time, any woman who visited Saudi Arabia had to have a designated male guardian or sponsor. The only way Shannon could make the trip was if Dan served in that role for her. Viewing that as demeaning, Shannon decided that wasn't going to work, either.

Then NASA came up with another idea. When Queen Elizabeth II had visited Saudi Arabia in 1979, the country had deemed her an "honorary man" so that she could freely dine and interact with the king. NASA approached Shannon about the possibility of using the same title so she could go on the trip. Shannon shut down that scheme pretty fast.

"Absolutely not," Shannon said. "I am who I am . . . There's no way I'm going to be an honorary male." The request had reminded her of something she'd read about years earlier, in 1970, when a New Zealand rugby team had traveled to South Africa, which was still under apartheid. The team had three players who were of Māori descent and one player of Samoan descent, and the color of their skin would normally have prohibited them from coming to the country. However, South Africa deemed them "honorary whites" as a loophole to let them enter. Shannon had thought that incredibly demeaning. "Look," she replied to the idea of designating her an *honorary man*, "this is so wrong. You are what you are."

Of course, when the other astronauts heard about this, the situation was ripe for ribbing. Mike Mullane found Shannon in her office one day and congratulated her on "having achieved the highest honor a woman could ever hope to achieve . . . to be designated an honorary man." Mike may have evolved some in his thinking, but his humor had yet to catch up. He avoided Shannon for a few weeks after making that comment.

Ultimately, NASA agreed that Shannon didn't have to go to Saudi Arabia and be an honorary male. Still, the decision had been a bummer for her. She loved to travel and always wanted to see new places. But she stuck to her principles and stayed home, while her crewmates jetted off to the Middle East without her.

Over in Saudi Arabia, once the men were off the plane they met up with Sultan. He looked around and noticed someone was missing. "Where's Shannon?" he asked.

"She's not here," John Fabian replied.

"Where is she? Why isn't she here?"

Without going into too many details, John tried to explain that Shannon had decided she didn't want to come. The news didn't sit well with the royal family, especially Sultan's mother, Sultana bint Turki Al Sudairi, who was throwing a lavish reception that she wanted Shannon to attend. Later, the crew learned secondhand that King Fahd placed a call to the White House and got President Reagan on the phone.

The next morning Shannon was sitting at her desk at work feeling sorry for herself that she wasn't off on a grand adventure like the rest of her crewmates. Then the phone rang. A NASA official was on the other end. He informed Shannon that she was leaving for Saudi Arabia in an hour.

Shannon thought about it for a minute. She couldn't afford to lose her job by saying no. Her kids were getting older, and she needed to save money for their education. But there were other complications.

"Well, I don't have a passport," Shannon replied.

The employee told Shannon that wasn't a problem and that everything would be provided for her. After the call ended, Shannon quickly got up from her desk, left JSC, went to the drugstore to get pain medication for her son who was home sick that day, and made her way to the house. While there, she called her husband, Michael, at work.

"I'm headed out to Saudi Arabia," Shannon told him, impressing on him the need to come home on time that day.

After putting three outfits in a suitcase, Shannon spotted a car pulling into her driveway. She got in the vehicle and the driver handed her a passport provided by NASA. Off they went to the airport, where Shannon was greeted by another person she'd never met before. The man walked her to a gate, and the two boarded a plane bound for New York City. No one asked who they were or what they were doing. *Hmm, this is interesting*, Shannon thought.

Three hours later, they were in New York and Shannon was told to get off the plane. Wondering what she'd do next, another man approached her and said, "Follow me." So Shannon followed and found herself on another plane bound for Saudi Arabia, with people conducting their daily prayers in the aisle. Completely clueless as to where in the country she was headed, she decided she'd go to sleep and simply get off the plane at the first stop.

By the time Shannon arrived in the Middle East, night had fallen. Wandering the arrival area a bit aimlessly and unsure of her exact location, she noticed another man approaching her. "I just got a call that you're showing up," he said. "Where are you headed?"

"Beats me!" Shannon answered.

The mysterious man, an American who seemed to work in the consulate, told Shannon he thought she was supposed to attend an event with Sultan's mother, Sultana. "How are you going to get there?" he asked.

"I don't know!" said Shannon. "I don't even know where I'm going."

They walked on together, passing by a Saudi man wearing a traditional white thobe who was about to board a small white Mitsubishi private jet. Shannon's companion called out to the Middle Eastern traveler.

"Hey, where are you going?"

"Riyadh," the man responded.

"I think she needs to go to Riyadh. Can she fly with you?"

"Sure!"

Soon, Shannon found herself high in the air, having a lovely conversation with this Saudi businessman she'd only met a few hours earlier. He had a great command of English, and when he learned Shannon was a pilot herself, he let her take the controls for a while. He even asked her if she wanted to land, but she declined out of an abundance of caution, since she hadn't been checked out in that specific plane before.

Finally, Shannon and her new friend touched down in a pitch-black Riyadh, and the mysterious journey continued when she found a black limousine awaiting her at the airport. Accepting her strange fate, she climbed inside and was transported to what she assumed was a hotel. There she put on a suit, one of the three outfits she'd brought, before she was whisked away to her next location.

At last, she arrived at an opulent palace where she found the rest of her crew and their spouses, who'd already been in Saudi Arabia for the past few days. Among her friends, she was able to relax slightly. She sat with them at a table during a banquet while women and men dressed in Saudi Arabian garments performed a presentation onstage and meals were brought out to the guests. Shannon sat there for hours, wondering exactly what was going on. But she was able to chitchat with her crew, and she learned later that Sultan's mother was happy that she was there, so she'd apparently fulfilled her duty to NASA.

The next day, after some much-needed rest at her hotel, Shannon wondered how exactly she'd get home. She had only about five dollars on her. But once again, when it was time to go, a car arrived for her and drove her to the airport. As she stepped aboard the plane, she'd been in Saudi Arabia for less than twenty-four hours.

Back in Houston, Shannon narrated to her daughters each leg of her whirlwind trip. They were shocked by her actions. "You mean you hitchhiked a ride on an airplane with a strange man that you didn't know?" they asked, aghast.

Shannon simply shrugged. It was in her nature to go with the flow. "Well, you know, you do what you got to do."

CHAPTER 17

Turning Point

I t was the end of 1985, and the mood among the astronauts at NASA couldn't have been brighter. All the TFNGs had finally flown their first flights, and a lucky few had already flown twice. Practically everyone had secured a new flight assignment, too. Sally, Judy, Anna, Rhea, and Kathy were all back in the crew rotation. Having just flown, Shannon didn't have a new flight, but it wouldn't be long until her next mission was finalized.

Everyone could feel the electricity in the air. The year 1985 had been a banner one for the Space Shuttle program. The astronauts had flown a total of nine flights that year, the most since *Columbia* had made its first flight. And 1986 promised to be even more epic. NASA was targeting up to twelve flights in the year ahead.

To meet the challenge, there were more astronauts in the corps than ever. NASA had brought new recruits into the program in 1980 and 1984, putting eleven women in the mix for crew assignments. Everyone could sense change in the air—recognizing that the Space Shuttle program was transitioning into a new, more vibrant phase.

Change was also in the air for Sally, though most of it was happening far away from NASA and Houston. It all began in 1984, when she gave a speech in Atlanta at the American Bar Association. It was just a routine speaking engagement, like hundreds of others she'd been asked to do. As usual, she asked Carolyn Huntoon to come along since she'd been such a supportive ally during the nightmarish press tour.

But Sally also saw an opportunity with this trip. She knew that Tam O'Shaughnessy, one of her old friends from tennis, lived in Atlanta. The two hadn't seen each other in years. While Sally had left the world of tennis in college, Tam had gone on to play professional women's tennis throughout the early 1970s, even competing at Wimbledon. When she retired from the sport, she earned her master's in biology from Georgia State and began teaching it to eighth graders in Atlanta. Tam and Sally had reconnected briefly while Sally was in graduate school in Stanford, and Sally had even invited Tam to the launch of STS-7. But they didn't really spend any time together in Florida. Sally was in quarantine and laser-focused on the mission. But the invitation had prompted the two to reestablish their friendship over the phone.

Sally rang up Tam before hopping on her flight to Atlanta to see if they could meet. "It's been too long," Sally told Tam. She suggested that Tam come to her talk, after which they could grab some food. Tam happily agreed. At dinner, while Carolyn sat at the bar, Sally and Tam reminisced about their tennis days. The banter felt so easy. It was as if no time had passed. When they paid for their meal, Sally knew she wanted to keep seeing her old friend again and again.

Throughout 1984 and well into 1985, Sally began making

periodic trips to Atlanta to see Tam. The two women would never do anything that elaborate together. They'd primarily take long walks around Tam's neighborhood, where they'd reminisce about their days in the junior tennis circuit or swap knowledge about physics and biology. "We just had fun talking about our dreams for the future and the past," Tam said. Every visit was just a conversation between friends. But whenever Tam would tell her friends that Sally was coming to town, they all noticed a little sparkle in her eye.

Meanwhile, back in Texas, Sally and Steve's relationship was deteriorating. Sally's frequent trips—either to see her family or Tam—were starting to take a toll. Steve recalled, "It wasn't uncommon at all for her just to say, 'Hey, I'm leaving town for the weekend.' And she would leave. It wasn't, 'Do you mind if I go see my parents?' or 'Do *you* want to go somewhere this weekend?' It was, 'I'm leaving.'" That feeling Steve had once had—that to Sally their relationship wasn't really a priority—was starting to grow again. Then, in mid-1985, Sally was assigned to her third flight. With Sally back in training, Sally and Steve barely saw each other.

One day that year, Sally escaped training in Houston to see Tam on just another routine trip. They took one of their typical long walks in Tam's neighborhood, grabbing some pizza slices before heading back to Tam's place. While sitting on the couch, Tam bent over to pet her dog. That's when she felt Sally's hand on her lower back. "Friends just don't do that," Tam said of Sally's touch. Surprised, Tam turned to Sally. "And I look back at her, and it was just . . . this moment," said Tam.

Tam saw the look in Sally's eyes. It was one of love. And Tam realized she felt it, too.

———————

IN LATE 1985, Judy was making her own reconnection with an old friend. She picked up the phone and dialed her old high school boyfriend Len Nahmi, now based in Los Angeles. In her hand, she held a gold Dunhill cigarette lighter. *His* lighter. He'd just sent it to her in an envelope in the hope that she'd fly it on her upcoming Shuttle mission.

Just like her friend Sally, Judy maintained good relationships with her exes—and she was happy to carry their tokens to space with her. She still called Michael Oldak on occasion, updating him on her major life events. On her first flight she carried dog tags that had belonged to Elissa Barker, the dog she'd shared with Michael when they were a couple in college. On that flight she'd also brought a gold watch that Len had given her. Later, he had it engraved with her mission's flight dates.

On the phone, Judy promised Len that his lighter would make it to space. They chatted for a bit about her plans for when she got back, and she told him her mandated press tour would probably bring her to LA at some point. When that happened, she promised they'd meet up. "I'll bring you your lighter," she said.

While the men of her past were still in her life, Judy was always one to move forward. She was no longer single. She'd started seeing a fellow astronaut named Frank Culbertson, a former naval pilot. Frank had been selected during the recent 1984 class selection, and the two had been going out regularly. It was her first steady relationship in some time. She'd dated off and on during her tenure at NASA, with rumors circulating that she'd gone out with a married astronaut. But ultimately, Judy was on her own most of the time.

During one New Year's Eve party at Dick Scobee's house, she found herself without anyone to kiss at midnight—the only person in the crowd without a partner. Later, she told Dick's wife, June, "I will never go to a New Year's event alone again."

Now she had someone to share those moments with—though, from the outside, people agreed that it seemed like Frank was more serious about the relationship than Judy was.

As always, Judy was most serious about training for her next flight. She'd been assigned to her second mission, STS-51-L, early in 1985. It was a flight filled with her TFNG classmates. Dick Scobee was her commander, and Ron McNair and Ellison Onizuka were her fellow mission specialists. Mike Smith, an astronaut from the 1980 class, would be piloting his first flight to space. They'd also have a payload specialist on board, Greg Jarvis, who'd been having some bad luck with his crew assignments. Originally slated to fly on Shannon's flight, he'd been bumped down the line when Rhea's flight absorbed Shannon's payload. Since Senator Garn stayed on Rhea's mission, Jarvis was assigned to another mission, STS-61-C, but congressman Bill Nelson, from Florida, would take his spot yet again. Finally, Greg was assigned to STS-51-L, during which he'd be conducting research on how weightlessness affected the movement of certain fluids.

It was already an eclectic bunch of astronauts, but Judy's flight would have yet another crew member on board—one quite distinct from the rest. The crew would be joined by the first teacher to fly to space, the one Reagan had vowed to find.

It'd been an exhaustive search, but eventually the selection was made. Out of more than eleven thousand applicants, NASA picked a social studies and history teacher from New Hampshire named Christa McAuliffe. Once Judy's crew got to space, Christa was going to perform school lessons, as well as conduct a few science experiments. Almost instantly, she'd become a media darling, appearing on all the morning shows and even *The Tonight Show* with Johnny Carson, who'd made women astronauts the butt of his jokes just a few years before. Some reporters even asked for her autograph. She stood in stark contrast to Judy, who actively shunned the media. Christa's easy demeanor and vibrant personality had made her a beloved public figure, even inspiring some envy among some of the other astronauts.

A few years back, Judy, along with plenty of other astronauts, had expressed concerns about flying non-scientists and non-astronauts on Shuttle flights. Mostly, she thought the influx of spacefaring lawmakers was "invading their space" and she'd mentioned to a colleague, "What are we going to do with these people?" But outwardly, she only showed her support for the newcomers and whatever NASA wanted her to do.

That was because she knew she wanted to dedicate her life to space. If that meant agreeing to NASA's PR stunts, so be it. She'd found her true passion working for the space program, and she wanted a front-row seat to the grand adventures NASA was rolling out. During a trip to D.C. to visit an old roommate, she sat on a nearby beach—one of the places she loved most—and told her friends that she was going to live on Mars someday. Coming from anyone else, it may have sounded silly, but anyone who knew Judy believed there was a chance she could make it happen.

Of course, Red Planet living was still many years off. For now, Judy would have to be content with her seat on the space shuttle *Challenger*, sharing the ride with this interesting newcomer. As Judy trained with Christa, she started to soften a bit to her presence and took the newbie astronaut under her wing. Because Christa's teaching focus had been social studies, she struggled with the math and science she needed to learn before boarding the Shuttle. Recognizing that, Judy met multiple times with Christa over coffee, giving her quick lessons and advice on how to make it through. Afterward, Christa always spoke highly of Judy in interviews, and she wrote to her parents that J.R. "especially had been very helpful."

As preparations for the flight advanced, Judy was looking forward to the amazing view she'd craved on her first mission. During launch and landing, she'd be sitting up in the flight deck, right behind the pilot and the commander. It was where Sally had sat during her first flight. Putting in the extra time with the simulators had paid off.

TOWARD THE END of 1985, Sally continued to escape Houston every few weeks, traveling to Atlanta to see Tam. By year's end, Sally and Tam had been seeing each other as more than just friends for the last five months. "It was kind of hard for a while, because she was going back and forth to basically, you know, live with Steve," Tam said.

But Sally was determined to end this limbo state. During one visit late in the year, she told Tam that it was time to come clean. She was going to tell Steve she wasn't in love with him anymore. And it wasn't just her marriage she was thinking about ending. She told Tam she wanted to stay at NASA through at least her third flight, but she knew that flight would likely be her last. She yearned to return to academia someday. She'd see space one more time, but after that, it was going to be time for a change.

Back in Houston, Steve was mostly oblivious to the nature of Sally's visits with Tam, but he could feel his wife's absence more and more. "It was tough," he recalled, "and I honestly remember thinking, I don't know how long I can put up with this." He kept thinking about how he'd made vows to persevere for better or worse. But this didn't feel like the marriage he'd signed up for.

New Year's Eve arrived, bringing an end to 1985. Sally knew that things were about to be very different for her in the year ahead as she contemplated her private life. But that last day of the year she pushed it from her mind and focused on another relationship. Her old crewmate, John Fabian, came to Sally at work with a request. It was his last day at NASA. His wife, Donna, had told him that his marriage had a "two-spaceflight limit." So after having flown with Sally and then Shannon, John was hanging up his astronaut wings. And on his last day, he wanted to close it out right. He asked Sally if she wanted to fly with him one last time.

The two hopped in a T-38 together, John in the front and Sally in the back, as was their usual configuration. They took off from Ellington and soared over the Gulf of Mexico. That afternoon, they did every kind of aerial trick one could imagine, flying upside down, barrel rolls, and more. They took turns assuming the controls, just as they'd always done once John knew he could trust his backseater. They stayed aloft all afternoon until the sun began to hang low over the horizon. In the air, they achieved a beautiful Zen moment— the year's perfect bookend.

Back on the ground, John embraced his backseater. "That's the way I wanted to finish," John told her, his voice slightly breaking. "That's the way I wanted to remember my time here." He thanked

Sally, knowing that for the rest of his life she'd always be part of that memory.

Sally knew that there was a final act in her near future too.

———————

IT WAS STARTING to feel like Judy's flight would never get off the ground. Originally, her mission was supposed to go up in November of 1985, but the target launch date got pushed back to January 22, 1986, thanks to issues with their main payload, a TDRS satellite. Then in December, NASA pushed the date back to January 23. They got all the way to January 22, when NASA delayed the launch again and then again.

Finally, on January 25, it seemed like they might be ready to fly the next day. But a weather check that night showed bad conditions throughout their launch window, and NASA opted to delay again just in case. Everyone thought it was an odd choice; usually NASA would press ahead with the countdown just in case. Weather in Central Florida was always unpredictable, and sure enough, there was nothing but clear skies on January 26, defying the launch forecast.

The frequent delays tested the crew's patience. And it didn't help that the flight *before* theirs, STS-61-C, had seemed like a nightmare. Hoot Gibson and Steve Hawley had both been on it, and it looked as if their Shuttle just did not want to fly. Judy watched in utter frustration as the crew climbed into the space shuttle *Columbia* a total of four times—once in December and three times in January—only to have to climb back out again when the launch was halted just a few seconds before takeoff. On their fifth try, Steve showed up at the launchpad wearing a Groucho Marx disguise. Steve figured if he was the one jinxing their launch attempts, they might finally fly if the Space Shuttle "didn't recognize him." That day they finally took off.

Judy loved the joke, and she even mentioned Steve's potential jinx when she first arrived at Cape Canaveral for her flight. "I am hoping that the affliction that Steve Hawley had from the 41-D mission—mission specialist of the delays—hasn't rubbed off on me," Judy joked on the tarmac at Kennedy Space Center. "And I think the guys behind me are hoping that it hasn't also. Otherwise, they might throw me off the flight."

But it was really starting to feel like a jinx *had* rubbed off on STS-51-L. The crew finally got a chance to head out to the launch-pad on January 27 and everyone had been strapped in for flight. At first, it really did feel like they were going to launch that day. Then the strangest problem arose. The ground crew couldn't get the hatch door to close properly. A fixture that needed to be removed to fully shut the door wouldn't come loose. Crew members tried everything to free up the piece of equipment, but it just wouldn't budge. Eventually, after nearly an hour and a half, they took a hack-saw to the faulty device and fixed the issue. But by that point it was too late. Strong winds had picked up near the pad, too intense for the Shuttle to perform a "return to launch site" abort if necessary. The winds never cleared, and they ran out of time in their launch window. A few hours later, the crew emerged slightly grumpy from the hatch that had ruined their takeoff.

The night of the scrub, Judy and the rest of her crew didn't re-ally believe they were going to launch the next day either. Weather forecasts predicted that temperatures were going to drop into the low twenties overnight, a rare chilly night for the normally balmy Florida coast. The general sense was that it was going to be way too cold to launch. The astronauts all crowded together in the crew quarters, perhaps more relaxed than usual with zero expectations they'd see space the next day.

Of course, all of this launch uncertainty was starting to weigh on the crew and their families. Ron McNair's brother and his pregnant wife ultimately decided to drive back to their home in Atlanta that day, because they couldn't wait indefinitely for the launch to take off. Judy's father and brother were still in Florida, though. They'd patiently waited through each of the delays, and they'd be on the top of the Launch Control Center when she took to the sky.

Overnight, the air circulating above Cape Canaveral plummeted as everyone had feared, dropping to just twenty-four degrees Fahrenheit. In an attempt to prevent the complex maze of piping around the Shuttle from freezing, NASA engineers pumped water through the Shuttle's fire suppression system. Drops of water trickled down the Shuttle and its launchpad, and as the air grew more frigid, the liquid began to freeze. Despite the best efforts of heaters placed throughout the launchpad, conical icicles hardened on the vehicle and the nearby service structure, resembling thousands of bared teeth, glistening in the dawning light.

Judy and the rest of the crew woke up at 6 a.m., almost half an hour earlier than they were supposed to. Because of the cold, NASA delayed the launch by an hour, allowing them some extra shut-eye. But they couldn't sleep in. They were all feeling antsy, ready to go. With ample time to spare, they shuffled into the dining room to eat breakfast, a time-honored tradition before every spaceflight. Judy sat at the large white dining table, which had a gaudy display of American flags and red and white roses at its center. On the menu were steak and eggs, two items that were always among the preflight choices. Feeling particularly hungry for protein that morning, Judy opted to eat two steaks as well as a generous portion of scrambled eggs.

After breakfast, the crew huddled in a teleconference room for their final weather briefing. Their fears from the night before turned out to be unfounded. Flight controllers at Johnson Space Center told them that, yes, it was cold and some ice had formed out on the launchpad. Crews had been out on the pad throughout the night, inspecting the ice thoroughly and attempting to remove some of the icicles. There'd been a concern that icicles might break off during launch, potentially damaging *Challenger*. But JSC didn't seem too concerned. There was no indication that the cold was a problem. The signs actually looked good. The crew left to suit up.

Clad in her light blue flight suit, Judy emerged from the gray doors of the building that housed the astronaut crew quarters and walked just behind and to the right of her commander, Dick Scobee. The chilly air hit them immediately, but the sky above was a beautiful crystal-clear blue. It seemed like a perfect day to launch. As Judy walked out to the Astrovan, she gave a slight low wave to the crowd gathered outside, her hand trailing behind as she turned the corner.

All the steps, all the motions, felt like second nature to Judy at this point. She'd done them all before, plenty of times for STS-41-D, and just the previous day before the scrub. There was the pull-up to the launchpad; this time they were launching from NASA's second main launch site, LC-39B, a near-exact replica of LC-39A. There was the climb in the elevator to 195 feet. There was the long walk across the suspended walkway to reach the Shuttle's cabin, though the crew had to walk carefully so as to not slip on the icy path. And there was the last look out at the Florida coast before entering the Shuttle. Admiring the clear blue sky, Dick Scobee reassured his crew.

"This is a beautiful day to fly," he said.

Judy hopped around to stay warm before entering the white room to get suited up with her headset and helmet. Next to her stood Christa, who'd be flying to space below her in the mid-deck. Before entering the cockpit, Judy turned to her mentee. "Well, next time I see you we'll be in space!"

Before long, Judy was back in the hard metal seat inside the cockpit with *Challenger*'s control panel and windows right in front of her. She ached to see the evolving view out the windows, but first came the necessary waiting game.

"Kind of cold this morning," Ellison said over the intercom from his seat next to Judy and behind Mike, the pilot.

Everyone knew Ellison hated being cold, so Mike took the moment to tease him about his position just off to the side in the cockpit, out of the view of the main windows.

"Up here, Ellison, the sun's shining in," Mike said over their headsets. "At least we've got the crew arranged right for people who like the warm and cool—got you out of the sun."

Flight controllers spoke into Judy's ear, asking her for a communications check. Feeling particularly amped that morning, she yelled in reply, "COWABUNGA!"

As the support personnel strapped Christa down below in the mid-deck, a gust of hot air suddenly blasted through the cabin. It was a gift from the closeout team to help heat up the freezing crew. While everyone whooped and cheered for the welcome warmth, Judy burst into laughter.

"It was right up my you-know-what," she said. "Ellison thought it was great."

The minutes ticked by, and they all felt that familiar misery of lying on their backs in a hard metal seat for hours on end, with Judy joking her "bun" was dying. Meanwhile, the threat of a scrub hung over their heads. The idea of getting out of the Shuttle and doing this same choreographed dance all over again seemed like an absolute nightmare. At T-minus 9 minutes, the countdown entered a planned hold—a pause that they all weren't sure they'd come out of.

"I hope we don't drive this down to the bitter end again today," Judy said, her excitement deflating.

But over the intercom, Dick heard from flight controllers; they'd reviewed the ice situation one last time and they were officially a go for launch. "All right!" he yelled, and Judy did a quick calculation of liftoff time: 11:38 a.m.

The minutes disappeared one by one, and with two minutes to go, Dick told his crew: "Welcome to space, guys."

At T-minus 6 seconds, the engines ignited as planned. Likely remembering her abort from STS-41-D, Judy waited to feel the rumble of the hardware, starting to generate more than one million pounds of thrust.

"All right," she said, relieved. As the engines reached full thrust, again Judy cried, "All *right!*"

And then, liftoff. Judy had been here before. The vibrations, the sounds, the forces on her body—she knew exactly what to expect this time and braced herself for each milestone. Her crewmates all kicked into action, sounding out each step on the checklist while celebrating their launch.

"Houston, *Challenger.* Roll program," Dick said aloud.

Mike continued with his jubilations. "Go, you mother!"

The Shuttle twirled onto its back, just as it did for every preceding flight.

It was Judy's turn. "LV, LH," she said fourteen seconds in, reciting the same acronym that Sally barely eked out during her first flight. But Judy was anything but nervous for this launch. "Shit hot!" she cried as *Challenger's* vibrations increased.

"There's Mach 1," Mike called out at forty seconds. They were officially traveling faster than the speed of sound. Just seventeen seconds later, they increased the thrust of their engines, giving them more power to reach their final orbit.

"Feel that mother go!" Mike yelled as the power increased. *"Woo-hoo!"*

Just over a minute into the flight, it was time to punch up the engines again, throttling them up to increase their power even more. Down in Mission Control, flight controllers gave *Challenger* the signal.

"*Challenger*, go at throttle up," Dick Covey, the CAPCOM on duty, said from the ground.

"Roger," Dick Scobee said. "Go at throttle up."

Outside the Space Shuttle, a bright flash appeared a mere three seconds later. Then a jolt. And in that brief moment, a realization that something was wrong.

"Uh-oh," Mike said.

CHAPTER 18

Closing a Chapter

Kathy found herself yawning in the Dallas airport the morning of January 28, 1986. She was on her way home to Houston, having just caught an early flight from San Jose, California. Back on the West Coast, she'd just had an exhausting week working at Lockheed, one of NASA's biggest contractors. It was work that was integral to her next Shuttle mission, which was going to carry something huge into space. The payload would be an observatory the size of a bus, one that would orbit the Earth while peering deep into the cosmos, snapping pictures of distant galaxies and nebulas. NASA was calling it the Hubble Space Telescope in honor of the famous astronomer Edwin Hubble, who'd been a key figure in the study of galaxies beyond the Milky Way. And it was going to be unlike any kind of spacecraft or satellite NASA had launched.

Perhaps the most intriguing aspect of Hubble was that it would get regular "tune-ups" while in space. NASA had the idea that astronauts could visit Hubble periodically and swap out its components, giving it upgrades over time to counteract the inevitable degradation of space. The term the agency used for Hubble was that it'd be "serviceable," invoking yet another car analogy. Astronauts riding on the Shuttle in the future would dock at Hubble and perform spacewalks around the telescope, acting as mechanics looking under Hubble's "hood."

Since Kathy and her crew were responsible for launching the telescope, they spent long exhaustive shifts with Hubble in its enormous white clean room in Sunnyvale, California, getting little sleep. Struggling to keep her eyes open at DFW, Kathy found a pay phone as she waited for her flight to Houston. She dialed her secretary to let her know she wouldn't be coming into work that day. Kathy had originally planned to go into the office when she landed that afternoon. But during the flight from California to Dallas, she passed out almost immediately from exhaustion. She realized she was just way too tired to be of any use in the office.

The phone rang a couple times before Kathy's secretary, Jessie, picked up.

"Hi," Kathy started. "I'm checking in." She quickly launched into an explanation about her exhausted state and that she wouldn't be making it to work that afternoon.

Kathy's words were met with a long, strange pause.

Then Jessie spoke, her voice shaking. "Didn't you hear what happened?"

IN A CONFERENCE room at an off-site facility, Rhea milled about with the crewmates for her next mission while someone found a TV nearby. They were meeting that day with a contractor, ready to undergo training for their upcoming flight. It was meant to be a routine day of training, mapping out exactly what they needed to do to prepare for Spacelab's launch.

They were still waiting on everyone to arrive when Rhea suggested they watch the countdown for STS-51-L. As they turned on the TV, one of Rhea's crewmates reminded everyone that the flight they were training for was originally supposed to fly that month. *This* flight was supposed to be *their* flight. Here Rhea was again, watching Judy take flight as she waited for her own to get off the ground.

With that reminder, Rhea watched *Challenger* climb into the sky with envy. She was more than ready for her crew to experience that same rush of launch again soon. For just over a minute, it was a picturesque launch, the small white-and-black *Challenger* rising against a vibrant blue sky, no clouds to be seen. The camera zoomed in tight on the Shuttle, the resolution of the video weakening as the vehicle's distance from the Earth grew.

Then suddenly, a jolt. A white cloud of flame and smoke engulfed the Shuttle in a fraction of a second. The camera switched to a wide shot, revealing a mass of plumes billowing out from where *Challenger* had once been. The Shuttle's solid rocket boosters awkwardly continued to climb into the sky, askew in their direction.

"See, there's booster separation," Rhea said, somewhat uncertain.

"No," said another of her crewmates. "It's too early."

SHANNON SAT IN Mission Control, listening to a training lecture. Having just flown a few months earlier, she hadn't been assigned to another flight yet. But she was already back in the generic training flow, part of which entailed doing ground support for an upcoming flight.

As she sat and listened to the speech about how to help as a CAPCOM, suddenly the lecturer stopped talking. Someone had just run into the control center, visibly shaken.

"There's been a problem with the Shuttle," he said.

Shocked, Shannon ran out of the room in search of the nearest television.

———

SITTING ON THE flight deck of the Shuttle mission simulator, Anna was positioned next to her crewmate Jim Buchli as she practiced the familiar motions of manipulating the robotic arm, controls she knew at that point like the back of her hand. Training was starting to heat up as Anna and her crew moved closer to their launch date. Their flight, STS-61-H, was slated to fly in just six weeks. Soon they'd be in the simulators practically every day.

Though she was focused on the current simulation, Anna kept wondering about STS-51-L. "Is the launch still on?" she asked one of the trainers again. Anna and her crew had a shared camaraderie with the 51-L crew. They'd all gotten their assignments at the same time and had gone out to celebrate together with the standard beer toast. Early in their training, their manifests had swapped around, as was typical, and the payload Anna's crew had originally been supposed to train on was now flying on *Challenger*. She'd listened to all the weather reports from the night before and was curious if the Shuttle was still going to fly. There hadn't been any indication that the launch was postponed.

During the simulation, Anna learned that the launch countdown had resumed after coming out of a hold at nine minutes. At that moment, she asked to freeze the simulation so they could go watch the launch in a conference room. Standing next to Jim, Anna watched the same sight Rhea was watching miles away.

Anna turned to her crewmate and they both agreed. Their simulation that day was over.

———————

SALLY SAT TUCKED inside her passenger seat in coach class. She was on yet another flight from Atlanta to Houston, a route she'd been taking every other week these days. The day before had been Tam's birthday and Sally made the journey to Georgia to celebrate with her. However, the trip meant that Sally was going to miss 51-L's launch. It was the first Shuttle takeoff she wouldn't see live. She watched all of the others on TV or saw them with her own eyes from Cape Canaveral. Or she'd been on the rocket itself.

But this exception was just all part of spaceflight becoming routine. Shuttle flights were ramping up, which meant you'd miss a flight or two sometimes.

At one point, a familiar ding broadcast throughout the cabin, indicating the captain was about to make an announcement. Likely anticipating a status update on their arrival time, Sally listened without expecting too much. But as the captain spoke, her stomach dropped.

Over the speaker, the captain announced there'd been an accident during the recent Space Shuttle launch. But he didn't have many details beyond that—just that there'd been a major problem. Sally's mind raced. *What happened? Were the astronauts okay?* She remembered that Judy was on this flight . . .

Sally rummaged through her bag and pulled out her NASA ID badge. She then unbuckled her seat and walked up to the door of the cockpit, flashing the badge at the cabin crew and explaining she was a NASA astronaut. Miraculously, the pilots let Sally inside the cockpit, giving her an extra headset so she could listen to the news coming in over air traffic control. They were only thirty minutes from Houston at this point, so the pilots were focused on landing. But soon updates started to trickle in.

The vehicle had completely broken apart, according to reports. Sally suddenly realized that Judy and all her friends were gone . . .

KATHY WENT TO NASA straight from the airport, no longer thinking about her sleep-deprived state. Her connecting flight had been a surrealist blur. She'd sat in shocked silence, surrounded by journalists all making their way to the city to cover one of the decade's biggest stories. She listened in quiet rage as they expressed what seemed like glee over the spectacle of it all. But through her anger, Kathy kept her mouth shut. *If you say a single word that makes them realize you're an astronaut, you're dead*, she thought. *They'll be all over you. No good is going to come of this.*

At JSC, Kathy found many of her colleagues floating around the office like ghosts. There she saw Sally, who also came straight from the airport. All Sally could think about was what the astronauts must have gone through during the breakup—what Judy must have gone through. Sally couldn't stop thinking about Judy; she'd been in the same seat Sally had occupied during her two flights.

"When I visualize what's going on in the cockpit during that accident, it's actually from that perspective," Sally would later say.

"From Judy's seat is where I picture, you know, what must have happened and what they . . . must have been going through."

None of the astronauts knew what they should do exactly. They all just wanted to be there, to help if needed. There was also some small comfort in being around people who were going through the same indescribable emotions.

With a knot in her stomach, Rhea rushed back to the JSC campus, pulling into the parking lot outside Building 4. She was about to dart inside when she saw Hoot waiting for her at the doors. From the building's upstairs window, he'd seen her pulling up and came down to meet her. Reunited, he wrapped her in his arms and together they cried out in the Texas daylight as visiting tourists walked by, confused by the scene and blissfully unaware of the tragedy unfolding on the East Coast.

The military astronauts had suffered losses like this before, watching a friend go in just the blink of an eye. It was devastating, but the feelings felt vaguely familiar. For the mission specialists what had just occurred was an entirely new kind of horror. The scale of it was unimaginable for everyone. They hadn't just lost *one* friend. Seven people were gone in an instant. Four of the original TFNGs. One of the Six.

Grief gripped their hearts and uncertainty hung heavy in the air. Underlying the sorrow, each astronaut thought the same thing, but they didn't dare say their fears out loud: Could this be the end of the Space Shuttle program?

That night, in the cover of darkness, Anna drove out to Ellington Field roughly six miles from JSC to await the arrival of NASA's training jets. On board was precious cargo: the families of the astronauts returning home to Houston after the worst day of their lives. Anna and other astronauts were there to greet them, to show their support far away from camera lenses and prying spectators. One by one, the families descended the steps from the airplane door and walked across the tarmac.

But Judy's family was not there.

———

FOR JUDY'S FATHER, Marvin, and her brother, Chuck, the *Challenger* launch had begun with a swift feeling of relief. They watched everything from the roof of the LCC, just as plenty of astronauts had done for previous launches. When the Space Shuttle cleared the launch tower, Marvin relaxed slightly, remembering just how dangerous those first few moments of liftoff could be. He thought back to Judy's terrifying pad abort from her first flight nearly two years before, and the anxiety that had gripped him in those uncertain first few moments. During that debacle, NASA personnel had hastily escorted him and the family into a huge area called the "ready room." Marvin knew why. It was a precaution—a way to prevent the crew's families from inadvertently watching their loved ones die in a fiery explosion on the launchpad.

But when *Challenger* soared into the sky on January 28, thoughts of that earlier abort quickly drifted from Marvin's mind. And for a little over a minute of flight, Judy's father was filled with nothing but pride for his space-bound daughter.

But then in a flash that enormous cloud ballooned outward from where *Challenger* had been, followed by an explosive noise on the ground a few moments later. Marvin watched the solid rocket boosters emerge from the billowing mass, creating what looked like a large Y in the sky. He searched frantically for any sign of the Shuttle breaking free of the cloud. But there was nothing.

Over the loudspeakers, NASA announcer Steve Nesbitt's voice emerged: "Flight controllers here looking very carefully at the situation. Obviously, a major malfunction."

In that moment, Marvin knew Judy was dead.

Next to him, an astronaut's children began to cry as the morbid realization sunk in. "Daddy! I want you, Daddy!" Everyone on the roof began to yell and sob, but it was the children's cries that would stick with Marvin for years to come.

And then, a sense of déjà vu. NASA personnel led by George quickly swooped up the families from the roof and escorted them into that same "ready room" Marvin had been in almost two years before. Except by that point, it was too late. They hadn't been able to prevent Marvin and the other families from seeing their loved ones die more than forty-six thousand feet in the air.

Huddled together, the families tried to process the utterly unthinkable. Their loved ones were gone. But even in those horrific moments, Marvin's warmth shined through. He noticed Lorna, Ellison Onizuka's wife, struggling with what had just occurred. Marvin went to her and asked her if she was okay. "Yes, I'm okay," she said, leaning against the wall. Then, as NASA engineers tried their best to explain what had just happened, the room plunged into darkness. Lorna had fainted, falling on some light switches on her way down.

THREE DAYS LATER, President Ronald Reagan stood at a wooden podium in front of thousands of mourners at Johnson Space Center, all sitting in white foldout chairs on one of the campus's open lawns. Most hung their heads as the president spoke, wiping away tears. As Reagan tried to memorialize the seven astronauts who'd been lost, the families of the *Challenger* astronauts tightly clutched one another for support. Some astronauts like Bob Crippen and Guy Gardner openly wept.

Missing from the Houston memorial, however, was the Resnik family. They'd flown back to Ohio the day *Challenger* exploded as the other families traveled back to Houston. Unable to make it to Texas, they'd stayed in Akron to attend a personal memorial service for Judy, held at Temple Israel. With roughly one thousand in attendance, it was a smaller service than the massive congregation at JSC, but it seemed to overflow as the attendees tried to pack inside the five-hundred-seat auditorium.

"Our Jewish tradition tells us that those who pioneer are partners with God," the rabbi overseeing the service told the crowd. "Judy heard an inner voice challenging her to climb higher. She heard a call to fly—to touch the sky. In that she excelled."

Marvin and Charles sat listening in the second row. Behind them a few rows back sat Judy's mother, Sarah. For her the day of the *Challenger* disaster differed wildly from that of the Resnik men. She hadn't been in Cape Canaveral for the flight. She hadn't even watched the launch live. She'd been in her apartment in Cleveland when someone called to tell her that there'd been a problem with the Shuttle her daughter was on. She'd immediately turned on the television to see *Challenger* reduced to a cloud of debris. It wasn't long before reporters started showing up at her front door and calling. Sarah declined most of their requests. Before being turned away, some journalists glimpsed newspaper clippings of Judy's accomplishments hanging on the walls of Sarah's home.

At Judy's Temple Israel memorial that day, the audience consisted of much more than family and childhood friends. Judy's NASA family was there, too. Both Sally and Kathy sat listening to the eulogies. They'd opted to fly to Akron instead of staying in Houston—to say goodbye to their friend and colleague.

Once the service ended, the mourners spilled outside into the open air and looked at the sky. Four NASA jets zoomed overhead. They flew in the missing man formation, spaced apart to make it seem as though one jet was absent. It was a traditional military salute that had been performed countless times. But this time, the person missing was a woman. Sally and Kathy huddled together with the women of the Resnik family as they extended their hands toward the jets, curling their fingers to make the ASL sign for "I love you." There, they saluted Judy one last time.

The next day, Sally was back home in Houston when her phone rang. Standing beside her as she picked up the receiver was her friend ABC correspondent Lynn Sherr and Lynn's husband, Larry, who'd come to town to cover the memorial. When Sally put the phone to her ear, NASA acting administrator William Graham's voice came through. He'd called to tell her that she was being appointed to a presidential commission, helmed by William Rogers, the former secretary of state under Richard Nixon. She and thirteen others would be responsible for deciphering what had caused *Challenger* to explode a little more than a minute into liftoff. She'd be the only current astronaut on the panel, though she'd be working with some of spaceflight's biggest names, including Neil Armstrong and Chuck Yeager. Controversial theoretical physicist Richard Feynman, who'd helped develop the atomic bomb, would also serve as a member.

It was an enormous ask. But Sally put down the phone and turned to her friends. "I need to do this," she said.

———————

A LITTLE OVER a week later, on February 10, Sally found herself sitting at a conference table with the other members of the commission—all male—in an office building next to the White House. It was a closed-door session—the third time they'd all met as an official group. Well, almost. Chuck Yeager failed to show up for the first two meetings. The excuse was that he'd been off "breaking another record." Chuck was in attendance for this meeting, though, only to announce that he'd have to leave again that night and that he wouldn't be back until March.

His presence wasn't really necessary though. These meetings were fact-gathering sessions to help the commissioners better understand the events surrounding the *Challenger* launch. Sally and her team began interviewing dozens of engineers and managers at NASA and at the Space Shuttle's various contractors, putting together the pieces of what went wrong that day. The sessions lasted hours and covered almost every topic imaginable related to the Shuttle system. And Sally was always ready to ask a pointed, clarifying question. Chuck didn't ask a single question during that lone meeting he attended. To Sally, he seemed bored.

The interviews had been illuminating. Just in the last week, the rapid-fire timeline of the accident began to take shape in the commissioners' minds. The culprit had been the Shuttle's right solid rocket booster, the same rocket the astronauts had always feared might tear their missions apart. More specifically, a thin piece of material inside the booster was to blame. The joints in the bottom of the booster—the thin rubber O-rings that sealed the big sections of the rocket together—had ruptured somehow, allowing the hot gases within the rocket to breach the outer casing. Like wildfire the flames swelled through the open rift in the rings, hitting the nearby giant orange external tank and causing a massive hydrogen leak. Then, a cascade of events occurred in less than a second, nearly imperceptible to the human eye. As the external tank weakened, the lower strut connecting the solid rocket booster to the tank began to come loose. As if on a swivel, the booster rotated quickly on its upper connecting strut, slamming into the external tank and releasing propellants that ignited in an all-consuming blaze.

It all happened in a flash before the crew even had time to react. And with no crew escape system, there was no way to flee from the disintegrating Shuttle. They could only sit in their seats as their lifeboat broke apart.

An even more ominous clue pointing to what ultimately befell *Challenger* was soon spotted in footage of the Shuttle during takeoff. A camera trained on the launchpad caught sight of a puff of black smoke spouting from the bottom of the right solid rocket booster right at launch. It was there for just an instant before it disappeared, but it was even more damning evidence against the O-rings. They'd failed at their main job of keeping the pieces sealed together.

Knowing all this was just part of the battle. The Rogers

Commission couldn't pinpoint exactly *why* this cataclysmic series of events took place. And what made things even more aggravating was that every step of the way, the press seemed to know more than NASA. Or, at least, the *New York Times* seemed to know everything. Before the commission even began to suspect the O-rings, the paper reported that power and pressure in the right solid rocket booster had suddenly dropped just before the explosion. Then, on February 5, the paper published a blockbuster story claiming that the solid rocket boosters weren't designed to handle temperatures below forty degrees Fahrenheit, pretty concerning given how cold it was the day *Challenger* launched. Later, Sally and her fellow commissioners asked mission managers if the cold was to blame but were told it was one of many factors being considered.

On February 9, the *New York Times* was back with another scoop. Apparently, the paper had obtained memos from someone at NASA showing that a staffer had previously voiced concerns about the possibility of the O-rings breaking one day, leading to a major catastrophe.

So NASA had *known* that the O-rings might be an issue. Blown away by this knowledge, the Rogers Commission decided to hold a closed-door meeting to gather more information and get to the bottom of the matter. But as the day wore on, the discussions were becoming exceptionally dull. A man by the name of Lawrence "Larry" Mulloy, who oversaw the solid rocket boosters at NASA's Marshall Space Flight Center, was going into excruciating detail about the booster's joints. Sally did her best to listen as the man droned on, flipping from one chart to the next.

After listening for what seemed like hours, the commission asked to take a short break. During the recess, Sally decided to return a few calls. One of them was to a reporter at the *Washington Times*. He'd heard a rumor that, before launch, one Shuttle contractor had expressed concern to NASA regarding the cold temperatures. Surprised, Sally couldn't confirm the rumor, but she decided to get to the bottom of it back in the session. When she returned to the meeting, her pink telephone slips still in hand, she got right to the point.

"Is there any internal correspondence on potential concern over the operation of the O-ring or the joint?" Sally asked. Her motive was to connect the dots. She wanted to see if maybe there was a connection between the O-rings failing and the cold temperatures—and if NASA had talked about it.

Larry said he couldn't think of anything but explained that there'd been a meeting the night before the launch at which engineers discussed the potential for problems due to the cold. But they decided to move forward with the mission. "We all concluded that there was no problem with the predicted temperatures," he said.

"I guess maybe what I'm asking is, we read in the *New York Times* about NASA internal memos where people within NASA were suggesting problems with erosion before, and I guess I am wondering whether similar memos exist relating to problems of launching with the O-rings at low temperatures," Sally said.

"I'm not aware of any such documents at Marshall," Larry replied.

The commissioners moved on, but soon afterward, a man sitting in one of the chairs against the back wall of the conference room raised his hand. No one really noticed. The people in the back weren't supposed to give any testimony at this session. They were simply there as "non-participants" in case the managers needed them to back up some statements. Meanwhile, Larry droned on with his charts.

Finally, the man with his hand up got out of his seat and walked to the main conference table, where the commissioners sat. Watching him approach, Mulloy turned and introduced the man.

"Mr. Chairman, Al McDonald from Morton Thiokol wanted to make a point," Larry said.

The members of the commission looked at this newcomer with some confusion.

"I wanted to say a point about the meeting," said Allan McDonald with a shaking voice. He was referencing the last-minute meeting Larry had mentioned earlier, the one the night before take-off at which everyone had seemingly agreed that the cold wouldn't be an issue for flight.

Then he dropped a bomb. Allan told everyone in that room that he and others at Morton Thiokol, the contractor responsible for manufacturing the solid rocket boosters, had originally recommended *not* launching in temperatures below fifty-three degrees Fahrenheit. They had data that indicated the O-rings lost resiliency the colder they became.

His admission was met with stunned silence.

"Could you stand up again and say that a little louder so we could hear it?" Chairman Rogers asked. "I'm not sure we all understood what you said."

As Allan elaborated about the late-night meetings and Morton Thiokol's initial recommendation, the mood in the room fundamentally shifted. It dawned on the commissioners that there *had* been concern about the cold. And engineers had recommended to NASA not to launch *Challenger*.

SALLY STARED AT an important piece of paper she held in her hands. The sheet, marked as an official NASA document, had two columns. The first listed temperatures in descending order. The second listed the resiliency of the O-rings as a function of those temperatures. To the most basic observer, the document showed that as the temperature decreased, the O-rings became stiffer and stiffer. The numbers held the key to what happened to *Challenger.* The O-rings only worked by staying flexible and filling in the gaps between the booster segments during flight, ensuring a tight seal. Stiffness meant the O-rings could become brittle and break. Stiffness meant death.

Someone associated with NASA had given Sally the document, but to this day, no one has publicly claimed responsibility for it. Steve always suspected it was one of NASA's various contractors, while Tam was certain it was an employee at JSC who knew of Sally's willingness to stand up for what was right. "Sally had this reputation at NASA and other places, just that she was a person of integrity and that she had a scientific mind," Tam said. "She couldn't be swayed by politics or that sort of thing."

Sally may have disliked the press, but she adhered to one of journalism's key tenets: Never reveal your source. She knew that revealing this person's name could lead to termination. And if she revealed the information herself, people might be able to trace it back to the original leaker. So she decided to quietly hand off the information. But to whom?

She'd already summed up her male comrades on the Rogers Commission in the short time she'd known them. Later, she even jotted down her observations in her notebook. Gene Covert: "tall, thin, a little gangly—drooping gray moustache, twinkle in his eye"; Albert "Bud" Wheelon: "talked in rewritten, edited sentences . . . eloquent." She didn't have much to say about Neil Armstrong, other than noticing his "slightly 'hick' suit" when they first met; she didn't know Chuck Yeager at all, seeing as how he never seemed to show up. And then there was Don Kutyna, a celebrated general who oversaw space systems for the air force. "Boyish, avid worker, good 'follower,'" Sally wrote in her notes, adding that he had chocolate in his briefcase. Sally liked Don from the start. She decided he was also a man of integrity and could be trusted to deliver the package.

One afternoon in February, Sally found herself walking down a hallway in the basement of the State Department with Don on her left. She realized this was the perfect moment. Without turning her head, Sally opened the notebook she was carrying and pulled out the piece of paper. She handed it to Don without uttering a word. As Don stopped to look at the paper, Sally picked up her stride and walked away, leaving her friend to ponder what he'd just received.

It took Don a few moments to process what he was reading, but he soon grasped the gravity of the numbers. Just like Sally, Don knew he needed to pass this information on to someone on the commission, preferably someone with fewer ties to the government. Don settled on a target accomplice: the eccentric physicist Richard Feynman. He'd understand the science, and he wouldn't be afraid to speak about it, loudly. The general decided to invite him for dinner to relay the news. After the meal, Don took the physicist out to his garage to see his 1973 Opel GT. He then removed the vehicle's carburetor, feigning that he was cleaning it.

That's when he got down to the reason for the visit. "Professor, these carburetors have O-rings in them," Don told him. "And when it gets cold, they leak." Richard didn't say a word, simply processing the information he'd been given. But in his mind, a light bulb had flicked on. *Of course they become stiff.* The two finished out the night without any more talk of O-rings.

On Tuesday, February 11, the Rogers Commission held another open meeting in Washington, D.C., with an abundance of audience members and camera crews to witness the investigation get underway. But before Richard arrived at the pageant, he'd gone to a hardware store and bought a C-clamp, some screwdrivers, and pliers. Taking a seat among two staggered rows of commissioners behind long tables, he then specifically requested a cup of ice water. Rather than sip it, he kept it on the table in front of him, and when the commissioners passed around a model of a solid rocket booster joint, Richard kept that too. He used his pliers to remove the O-ring from inside the seal and put it in his ice water. He waited a while as the O-ring grew cold before he reached out and pressed the button in front of him to turn his microphone on.

Addressing Lawrence Mulloy, who had earlier downplayed the role of temperatures in the O-ring failure, Richard pulled out the O-ring.

"I took this stuff that I got out of your seal and I put it in ice water, and I discovered that when you put some pressure on it for a while and then undo it, it doesn't stretch back," he said. "It stays the same dimension." He'd just demonstrated to a stunned room what NASA's managers had been reluctant to admit: the O-rings lose their flexibility when they get cold.

"I believe that has some significance to our problem," Richard concluded.

It was a small science experiment that Richard Feynman would become famous for. But it all began with a game of telephone in which Sally Ride was a key participant.

SALLY MAY HAVE helped crack the case of the O-rings, but the four remaining women in the Six all did their part in the wake of *Challenger*, whatever it took to keep the space program afloat. At first, Anna and her crew went through the familiar motions of simulations. No one quite knew how the accident would affect the program, so they continued to train in case the Shuttles did start flying again soon. That way the astronauts and ground crews could keep up their proficiencies.

But it became clear that none of the Shuttles would see space anytime in the near future. At that point Anna, Shannon, and the other astronauts were assigned tedious work, going through every checklist and every document related to past Shuttle flights. Their job was to streamline all the procedures and manifests that came before. Often, NASA had listed multiple ways of doing certain tasks or held various repeat documents, and the hope was that making this more efficient might decrease risk. "It was a lot of long, long hours, sitting, going line by line, through every single piece of documentation," said Shannon. And they did it for *months*.

Since Kathy had been immersed in the new Hubble program before the accident, she didn't take on an official role in the accident investigation. Rather, she embraced an unofficial role of support, doing her best to visit the families of the *Challenger* crew and help carry any loads they struggled to lift. She continued to make the rounds, even as the initial crowds of mourners and well-wishers slowly disappeared. Of all the families and relatives to lose loved ones, Kathy felt particularly drawn to June Scobee, whose husband had been *Challenger*'s commander. Kathy had been close to June and Dick before the Shuttle took flight.

Now in the months after the explosion, Kathy often found herself sitting at the kitchen table in June's house, listening to the grieving widow as she talked about her ideas for honoring Dick's legacy and the lives of his crewmates. Taking shape in June's mind was some kind of "living memorial" for the *Challenger* crew. She told Kathy she didn't want to just slap some names on buildings and schools. She wanted something that would continue to spark the imaginations of children for decades to come. An education initiative might do the trick.

Kathy offered her thoughts, but mostly her open ears, helping June channel her grief into a concrete plan of action.

Before the crew's legacy could be upheld, though, they had to be found. At first, it wasn't clear if there'd be anyone left to recover. Everyone at NASA assumed the crew died instantly in the blast, vaporized in seconds. How could it be any other way? The footage had seemed unambiguous. Though such an end was tragic, there was some small comfort in the notion that their deaths had been quick.

But more and more folks analyzed the footage in the weeks following the explosion, playing the tape on a loop while meticulously scanning the video for the smallest hint of details that might have been missed. Over time, eagle-eyed analysts made a grim discovery. Buried within the grainy footage of the accident, they found a small white speck trailing away from *Challenger*'s destruction, heading toward the sea. It was the crew cabin.

Thus, one of the most expansive underwater search efforts in history began, with NASA employing the efforts of the navy, the coast guard, the air force, and various agency contractors. Dozens of ships, aircraft, and submarines—along with thousands of rescue personnel—worked together to scour the depths of the ocean floor in the hopes of finding where the crew of seven returned to Earth. Nearly six weeks elapsed at a cost to NASA of up to $1 million a week, before divers onboard the USS *Preserver* found the remains of the cabin in the murky depths.

In its final resting place, the cabin resembled a crumpled-up sheet of aluminum foil. "It looked as though the cabin had been blown up by a bomb, then the pieces swept together into a pile on the ocean floor," one newspaper described. Housed within the mass were what remained of the crew. Judy's remains were some of the first to be freed. Divers recovered strands of her famous hair, as well as the gold necklace she'd been wearing. Every astronaut who knew Judy recognized the necklace immediately. She wore it nearly every day and even took it to space with her on her first flight.

Understanding the sensitivity of the operation, NASA hoped to keep its crew recovery efforts a secret. But with so many journalists and members of the public hungry for details about what had happened to the *Challenger* crew, it was next to impossible to cloak NASA's salvaging efforts. Photographers lined the coast to spot the ins and outs of ships canvassing the waters, while journalists bought handheld radios to listen in on the frequencies the searchers used for communicating their discoveries. Under the cloak of darkness one night, the initial remains were shipped to port. But someone, hoping to show respect, draped the containers carrying the crew in American flags. The press snapped some pictures and the world knew that the *Challenger* astronauts had been found. It would take three months to salvage everything.

All the debris and human remains that NASA recovered were initially stored in Hangar L—a large open warehouse at KSC. The hangar served as a macabre museum, showcasing the bits and pieces of *Challenger* that NASA managed to fish out of the water. And it also served as the crew's temporary home. NASA engineers and astronauts came to the hangar to identify the wreckage.

One day after the search ended, Rhea was asked to retrieve Mike Smith's flight suit from Hangar L and clean it up, so that it could be used for his upcoming funeral. Rhea was given the tattered and torn suit, which reeked of an "industrial-style stink." She found some detergent and got to work.

The more she scrubbed, the cleaner the suit became. But it wasn't enough for Rhea. As she rubbed her hands raw cleaning the suit, tears began streaming down her face. No matter what she did, the suit wasn't clean enough for her. And she wondered aloud why this had to be happening.

———

IN THE SUMMER of 1986, NASA concluded its investigation into the final moments of the *Challenger* crew. And its findings dashed the hopes of everyone who'd prayed that the astronauts were immediately killed when the Shuttle disintegrated. No, they'd survived the breakup, investigators found. The evidence for that could be found in the astronauts' PEAPs, or personal egress air packs. They were boxes connected to each crew member that could provide them with six minutes of oxygen in the case of a catastrophe. Three of the four PEAPs that were recovered had been turned on. It was a sign that the astronauts had been both alive and conscious after the failure, actively making the decision to flip those switches in the precious moments following the Shuttle's destruction.

To this day, no one knows if the crew was conscious when they hit the water. It's possible that at some point during the two-and-a-half-minute plunge into the Atlantic that the cabin lost pressure, causing the inhabitants to pass out at the dizzyingly high altitudes. They weren't wearing pressure suits, just long-sleeved jumpsuits and helmets—little protection from the thin atmosphere surrounding them. Such a fate would have mercifully spared them the experience of slamming into the ocean's surface at roughly two hundred miles per hour, an impact equivalent to a car ramming headfirst into a concrete wall. But it *is* possible the cabin lost pressure only slowly, keeping the crew alert and aware as the cabin lurched downward and picking up a painful amount of g-forces until the ocean rushed at them in the cockpit windows.

Of all the PEAPs that were activated, the most interesting one found was that of pilot Mike Smith. The switch for his PEAP was located on the back of his chair. That meant that either Ellison or Judy, both of whom sat behind Mike, had switched it on in those harrowing first few seconds of disaster. While it's more likely that Ellison flipped the switch (he'd been trained to do so given his position in the cabin), Judy had equal reach to activate the machine.

Even in those chaotic few seconds of terror, fully knowing what lay ahead, it's possible that the last thing Judy did before she died was attempt to save the life of one of her crewmates.

IT WAS NEARING the end of the investigation, and the Rogers Commission was debating which color their leather-bound report should be. The two finalists were maroon and blue. Robert Hotz, an editor at *Aviation Week* and one of the commission members, gave a speech on the "elegance" of maroon, while describing why blue was "inadequate." Ultimately, it was put to a vote. Nearly everyone opted for maroon but two favored blue. Then Hotz turned to Sally. Perhaps she could be the color coordinator, he suggested, as the lone woman on the commission.

Annoyed by his comment, she deferred to Don. Later, Don told Sally he couldn't believe Robert said that to her.

The report showed up the day before it was supposed to be submitted, bound in blue leather, and Sally got the blame for it.

The color of the report was, of course, irrelevant. It was the contents of the massive 450-page document that would change NASA forever. In those pages, the members of the commission detailed NASA's troubled history with the solid rocket booster's O-rings. Before *Challenger*, up to seven flights had experienced some form of O-ring explosion—including Judy's first flight—many of which occurred at colder temperatures. It was a stinging indictment of NASA's safety procedures and how the agency's pressure to increase the Space Shuttle flight rate may have contributed to corners being cut. The report recommended almost a complete overhaul of how NASA did business, even suggesting that satellite operators begin launching their payloads on expendable uncrewed rockets again.

The commission turned in its report on June 9, 1986, after a grueling four months of work. All the while, Sally remained a model of professionalism, her demeanor belying any underlying problems in her personal life. But her life was fundamentally shifting. After plenty of uncertainty about the future of her and Steve's relationship, she went ahead and made the decision for them. One day in late 1985 she told him she didn't want to be married anymore. "I don't know what would have happened had she not made the first move on that," Steve said. But at the time, there wasn't much to be done to acknowledge the separation. The two were intricately tied into their respective jobs after *Challenger*. So they lived together as roommates until they could figure out what to do. All the while, they kept their news a secret as they toiled away at work.

With the Rogers Commission report now complete, Sally knew it was time to start the next chapter of her life. But she wasn't ready to leave NASA just yet. She knew it was a bad time to leave, as the wounds of *Challenger* hadn't fully healed. So she made an offer to the NASA administrator. She'd work out of headquarters in Washington, D.C. And with that decision, NASA's first American woman in space was no longer in the astronaut corps.

There was still one final project to be completed, though. At the administrator's request, Sally formed a planning group aimed at coming up with a bold and extensive framework for NASA's future. With the Space Shuttle fleet temporarily grounded, the agency found itself in need of direction. Sally and her team spent nearly a year developing an intricate new path for NASA to take. In 1987, they turned in their report, which informally became known as the Ride Report, outlining four major initiatives that NASA should focus on.

The most ambitious goal was sending humans to Mars, after creating an outpost on the lunar surface—two major enterprises that NASA still aspires to achieve. The second most important goal was to explore the solar system robotically, an objective that is on its way to being fully met thanks to the space agency's robust planetary science program. But there was also an item on the list that was intriguingly fresh. It was called "Mission to Planet Earth," and it called on NASA to launch a suite of satellites to observe Earth from space—to catalog how the planet's weather and climate changed over time.

Sally's life-altering moment gazing out at the fuzzy blue atmosphere of Earth had manifested in a big way. She knew Earth's atmosphere needed protecting, and she was calling on NASA—an agency yearning to strike out beyond humans' home planet—to turn its sights inward. Mission to Planet Earth was a controversial suggestion that NASA considered burying at first, and Sally thought it might lead to her being fired.

But she decided to be the master of her own destiny. With her report turned in, Sally retired from NASA in 1987 and packed her bags for California, where she was set to begin her new post-space academic life at Stanford. It was time to start something new, though her trips to and from Atlanta would continue. For Sally, it was a necessary shift. But for America, it was a challenge. After losing its second woman to fly to space, it had now lost its first. The woman who'd won plaudits as a pioneer of space exploration was no longer an astronaut.

AS SALLY PREPARED to say her goodbyes, Anna Fisher sat at a conference table among a gaggle of NASA officials and managers. Nearly a decade after she'd interviewed in front of the astronaut selection board, she was about to experience the whole process again, but this time *she* would be the one making the selection. She enjoyed being on the opposite side of the table, noting that it was way less stressful to be the one choosing the next class.

It certainly was an interesting time to be selecting new astronauts. NASA was still gearing up for its return to flight with the Space Shuttle—at least a year away. With so much uncertainty surrounding NASA's future some questioned the need to hire more spacefarers. But others figured the agency needed to keep numbers up. They didn't know how many astronauts would leave in the wake of the disaster (or that Sally was on her way out the door).

Anna sat through interview after interview, realizing just how much a person's first impression stuck in her mind. Choosing among so many skilled applicants was an enormous task, but just as the 1978 selection committee had done, she and her colleagues narrowed the list to fifteen finalists. Among the group, two more women would join the corps. One of those women was named Mae Jemison. With her selection, NASA would be making history yet again, welcoming the first Black woman to the ranks of spacefarers.

It had taken decades, but the agency had finally learned an important lesson: courage and perseverance in the most pressure-filled situations aren't traits that are exclusive to a single gender or race.

B ack in late August 1978, the year that the Six made their debut as astronaut trainees, a woman by the name of Eileen Collins was in the midst of pilot training at Vance Air Force Base in Oklahoma when she heard that a group of astronauts would be visiting in the coming days to undergo parachute survival training. Among the arrivals were several mission specialist astronauts from a recently chosen group of thirty-five. And on the list of attendees were the first women astronauts that NASA had ever picked.

For Eileen, it was inspiring news. She never saw the astronauts during their time at the base, but knowing they were nearby gave her a renewed sense of purpose.

"Just the fact that they were on the same base I was on was very exciting," Eileen said years later. "I thought, *This is something that I'm going to do someday.*"

Flying had been her first real passion, ever since she was a little girl living in Elmira, New York—known as the "Soaring Capital of America"—where she watched a dazzling array of gliders fly overhead every day above nearby Harris Hill. But she'd also been incubating the idea of becoming an astronaut since the fourth grade when she read an article in *Junior Scholastic* magazine about the astronauts in the Gemini program. She, of course, noticed the lack of women, but figured she'd somehow find a way to become a "lady astronaut."

Doors seemed to open for Eileen in 1976. That was the year the air force reversed its ban on women flying jets and accepted its first ten women pilots into training—a group Eileen tracked with keen interest. After graduating from college two years later, she joined the air force's flight training program, putting herself on a path that would eventually lead to NASA.

But unlike the women astronauts who came before her, Eileen would be chosen not as a mission specialist but as a Space Shuttle *pilot*.

THERE WAS A time when it was unclear if anyone—let alone women—would be able to fly to space again. Fears abounded that the *Challenger* accident would spell the end of the Space Shuttle program and NASA's pursuit of space exploration. But the agency persevered through what was then one of—if not the—most difficult tests in its history.

Answering the wake-up call that the *Challenger* tragedy represented, NASA completely reevaluated its safety procedures, and engineers redesigned the Shuttle to make it safer for crews once they were again permitted to step aboard. The solid rocket boosters underwent significant design changes to add redundancy and increase safety. The pressure to increase flight rates eased. The country abandoned plans to launch the Space Shuttle out of California, choosing instead to focus on Florida launches—and NASA dropped all plans to fly more civilians, including an unpicked journalist. And the light blue flight suits that all Shuttle astronauts had worn up to that point were swapped for new orange pressure suits—ones designed to keep crew members alive for a short time in case the vehicle suddenly lost oxygen and pressure.

NASA was doing its best to learn from its mistakes. Still, the agency needed to make things right for those who'd been lost.

Judy's father, Marvin, wound up turning to a significant figure from her past, Michael Oldak, for legal help as the family sought to get a settlement from NASA. "I said if they wanted, I'd be willing to represent the family," Michael said. "Judy put me through law school. I'll do this for them." But as Michael recalled, there was reluctance to provide the Resnik family with a cash settlement comparable to those the other families would receive, since Judy at the time of her death was both unmarried and without children. The differing amounts led to a contentious back-and-forth with NASA, and at one point, Michael said that all communication with the agency temporarily ceased.

So he decided to do some heavy recruiting. He hired the D.C.-based law firm Williams & Connolly, a litigation powerhouse. "I told Morton Thiokol, you can settle with me or fight with them," Michael said. In the end, Judy's family received a settlement that Michael felt was appropriate.

The tragedy's aftermath may have left many feeling unmoored, but a light did emerge from the darkness. June Scobee's dream of a living memorial for the *Challenger* crew came to fruition in the form of a nonprofit, the Challenger Center, which aims to use the legacy of the astronauts to inspire children to pursue careers in STEM.

Marvin would serve on the advisory council before his death; Judy's brother, Charles, now does so in his father's stead.

The *Challenger* accident powerfully reverberated across the personal and professional lives of the rest of the Six. It meant, for example, that Anna's rescue mission in space would be the one and only spaceflight she'd take. Though she was scheduled to fly on a second flight, the uncertainty surrounding all the crew assignments after *Challenger* ultimately led her to take time off from NASA. In 1989, the same year her second daughter, Kara, was born, Anna began a seven-year leave of absence to focus on raising her family. She returned in January 1996 to a very different Space Shuttle program. "Coming back after seven years was probably the hardest thing I've ever done," Anna said years later.

She ultimately re-found her footing, becoming an instrumental figure in the early development of the International Space Station. At one point Anna thought she might even fly again, finding herself back in the lineup for another flight. But once again tragedy kept her grounded. On February 1, 2003, the Space Shuttle *Columbia* broke apart as it reentered the atmosphere over Texas, killing all seven astronauts on board. The loss put flights on pause, and Anna ultimately retired from NASA in 2017 after having worked at the agency for nearly forty years. "I was disappointed to not fly again, but I also realized how lucky I was," she said, arguing that to be one of the few astronauts to have ever flown to space "is a truly humbling experience."

Rhea Seddon remained at NASA and would go on to fly twice more on the space shuttle *Columbia*. She'd fly not one, but two Spacelab missions—the same flight she'd dreamed of conducting since joining the Shuttle program. On her final flight in 1993, Rhea was the payload commander, in charge of all science experiments performed during the mission. She had two more children with Hoot, who'd wind up flying a total of five Shuttle missions at NASA and serve as the commander of the first Shuttle flight to dock with the Russian Mir space station.

Rhea retired from NASA in 1997, just two years after her daughter Emilee was born. The couple moved to Rhea's home state of Tennessee, where she worked as the assistant chief medical officer of the Vanderbilt Medical Group in Nashville for more than a decade. Eventually, they made their way to Murfreesboro, bringing Rhea back to the small hometown she knew as a child.

Kathy also stayed at NASA and flew two more Shuttle flights. She'd eventually fly on the mission that deployed the groundbreaking Hubble Space Telescope—the assignment she'd been training for before the *Challenger* accident. Also among the crew was Steve Hawley, marking the third of his total five Shuttle flights. It was a pioneering flight for NASA, and the Hubble Space Telescope continues to make observations of the universe to this day.

Kathy's involvement in government service wasn't restricted to NASA. In 1988, she was commissioned to the U.S. Naval Reserve as an oceanographer, ranked lieutenant commander. And in 1993, she was confirmed by Congress as the chief scientist of the National Oceanic and Atmospheric Administration. Years later, in 2014, she moved up to the role of NOAA administrator. She also never gave up on her dream of exploring the Earth's deepest waters. She'd take submersibles on multiple trips, including a voyage in 2020 to Challenger Deep—the lowest known location on the planet. She became the first woman to make the trip, and she's the only person to have walked in space and traveled to the sea's deepest depths.

Shannon may have been the last of the Six to fly, but she'd go on to bank the most spaceflight time of all of them combined. After the *Challenger* disaster, she flew to space four more times, traveling on the space shuttles *Atlantis* and *Columbia*. Her most notable trip to orbit took place in 1996, when she flew aboard the *Atlantis* for the STS-76 mission and visited the Mir space station. She didn't return to Earth with the rest of the crew but remained on the Mir for the next six months, spending most of that time with two cosmonauts, Yuri Onufrienko and Yuri Usachov. When she returned to Earth, she held the record for the longest time spent in space by an American and by any woman—records she would hold until 2007.

For Shannon, it was both a deeply fulfilling and challenging experience. Neither Yuri spoke any English, and she had to learn Russian in the year she trained for the flight. There were also minor complications. At one point, her daughters had given her a science fiction novel to read to pass the time. Shannon devoured it, only to find that the book ended on a major cliffhanger, leaving her in the dark about what happened next. Without a bookstore nearby, she stewed over how the plot would resolve itself. However, Shannon said that her daughter Kawai "made sure that the second edition was on the next Progress" cargo flight.

Despite the necessary delays in receiving care packages, Shannon learned that she preferred *living* in space over conducting short weeklong visits.

As for Sally, she returned to the world of academia as she'd planned, first taking a fellowship at Stanford, before landing a more permanent role as a physics professor at the University of California,

San Diego. But the biggest focus of her post-NASA life centered on public outreach, especially inventing ways to engage children. "I would say that most of her friends were shocked," Tam said. "Sally? Kids? It had never been her focus, ever." During her time giving speeches as America's first woman in space, though, she realized it was children she loved speaking with the most—and she also thought they asked the best questions.

Sally organized outreach for two major projects, EarthKAM and MoonKAM, which give children an opportunity to take pictures from cameras in low Earth orbit and in orbit around the moon. But by far her biggest passion was building a company she created with Tam called Sally Ride Science, a nonprofit associated with UC San Diego. The program focuses on using space as a way to inspire young children, notably girls, to pursue careers in science and math. Sally also maintained a relationship with NASA, helping with the investigation that followed the *Columbia* disaster. She was asked twice to be NASA's administrator, but always turned down the job.

Throughout their decades together, Sally and Tam never made their relationship public but remained steadfast partners—in both business and love. Then, in 2011, the unthinkable occurred. Sally was diagnosed with pancreatic cancer after doctors found a golf ball–sized tumor in her abdomen. She underwent intensive treatment and surgery, but on July 23, 2012, she died at her home in California surrounded by loved ones. After her passing, Tam posted an obituary to the Sally Ride Science website, revealing herself as Sally's "partner of 27 years."

The news went viral since it effectively revealed Sally to be the first known LGBTQ astronaut. A mixture of opinions flooded the internet, with some decrying Tam and Sally's decision to stay silent all those years. But Tam said that the outpouring of love and support easily drowned out the negativity.

"I have heard from people—friends, journalists, and other folks—who said that hearing that Sally and I were a couple made a huge difference in their coming out and being who they really are," Tam said. "So that's just fantastic."

———

IN THE YEARS after the Six left their mark on NASA and the world, the ranks of women in the astronaut corps continued to swell. Every single astronaut class since the historic one of 1978 has included women, with more recent selections comprising a nearly equal number of men and women candidates. And notable firsts have followed. Mae Jemison, selected by the committee Anna had been on in 1987, became the first Black woman astronaut and first woman of color in space in 1992 when she flew aboard the space shuttle *Endeavour*. Eileen Collins made history in 1995 when she became the first woman to pilot the Space Shuttle. She also became the first female commander of a Shuttle mission in 1999. And astronaut Peggy Whitson became the first female commander of the International Space Station in 2007.

Recalling the Six, Eileen said, "They didn't have any women [astronaut] role models. They were doing it for the first time. For those of us who followed, we had the role models. So that made us more comfortable, more confident, and more welcome."

But even as NASA and the world have made strides, there's still a long way to go to reach total parity. Of the more than six hundred people who've gone to space, less than one-sixth have been women (a ratio that hopefully will become out of date soon after this book's publication). And for women of color, the representation is even more pitiful. Only five Black women have gone to orbit. The first Hispanic woman, Ellen Ochoa, flew to orbit in 1993, and only one other has followed. And the first woman of Native American descent, Nicole Aunapu Mann, only went to space recently in 2022.

Meanwhile, NASA's decisions about women in space from decades earlier still reverberate today. In March 2019, NASA announced plans for two women to perform the first all-female spacewalk outside the International Space Station, which would be directed by a woman flight controller on the ground. But just before the astronauts were set to step outside the airlock, one of the women, Anne McClain, realized that a medium-sized suit would fit her best. Only one medium was available on the space station at the time; the rest were larges. So Anne chose to bow out, giving the sole medium to her female crewmate while a male astronaut took one of the larges.

The rescheduling caused an uproar among those looking forward to witnessing the historic moment. It also shined a light on NASA's space-suit decisions, which began when the first six women joined the agency all those years ago. Building and maintaining smaller-sized space suits had simply not been given higher priority over the years, preventing many smaller women from performing spacewalks altogether. Amid the 2019 space-suit kerfuffle, astronaut Jessica Meir commented, "I think that when people try to understand why we have the system we have—when you have technology that was developed and hardware that takes a long time to be proven and tested and make its way to spaceflight—sometimes the effects of those decisions made back in the '70s carry over for decades to come."

NASA didn't take long to make things right, though. Jessica

Meir, along with her best friend and fellow astronaut Christina Koch, finally did make history in October 2019 when they both donned medium space suits and stepped out of the airlock of the International Space Station to perform repairs and swap out batteries on the vehicle's exterior.

In the past couple of years, space travel has begun to look very different from what it looked like when the Six came to NASA. Originally, human space exploration was squarely the domain of NASA and government agencies, but now, as the commercial space industry blossoms and strengthens, there are more opportunities for women and people of color to find their way to space without federal approval. And an astronaut's résumé need no longer include any piloting or science experience at all. Companies like Blue Origin and Virgin Galactic are promising quick trips to space for customers who can pay their way. And they've even allowed for a few historical corrections. Wally Funk, one of the thirteen women who passed Randy Lovelace's tests back in 1961, finally got a chance to see space in 2021, when she flew on board one of Blue Origin's rockets to the edge of space and back.

Meanwhile, companies like SpaceX—which has become a major NASA partner—are going even further by enabling average citizens to fly all the way to orbit. In September 2021, a billionaire-backed trip on one of SpaceX's vehicles allowed two women, childhood cancer survivor Hayley Arceneaux and Professor Sian Proctor, to have the opportunity to lap the Earth for three days.

And as NASA moves forward, it's keeping women top of mind. In 2017, the agency formally announced the creation of the Artemis program, an effort to send humans back to the moon for the first time in more than half a century. Artemis, named after the goddess in Greek mythology who is Apollo's twin, will be striving to send the first woman and the first person of color to the lunar surface.

In April of 2023, NASA took a major step toward fulfilling that promise. Christina Koch, one half of the first all female spacewalk, and Victor Glover were assigned to the first crewed Artemis mission to loop around the moon. When they fly, they'll become the first woman and first person of color to travel to deep space respectively.

And after them, the world will learn of the next woman to make history in space. Among the current members of NASA's corps is the first woman who will walk on the moon. All she's waiting on is her selection.

On September 17, 2012, twenty-two former and current women astronauts, along with NASA's first female director at Johnson Space Center, reunited in Houston to honor Sally Ride after her passing earlier that summer. Seated (from left to right): Carolyn Huntoon, Ellen Baker, Mary Cleave, Rhea Seddon, Anna Fisher, Shannon Lucid, Ellen Ochoa, and Sandy Magnus. Standing (from left to right): Jeanette Epps, Mary Ellen Weber, Marsha Ivins, Tracy Caldwell Dyson, Bonnie Dunbar, Tammy Jernigan, Cady Coleman, Janet Kavandi, Serena Auñón-Chancellor, Kate Rubins, Stephanie Wilson, Dottie Metcalf-Lindenburger, Megan McArthur, Karen Nyberg, and Lisa Nowak. NASA

AUTHOR'S NOTE

This book is the culmination of more than one hundred hours of interviews (mostly over Zoom during the COVID-19 pandemic) conducted between 2020 and 2022. Among the people I was grateful to talk to were Augusta Gonzalez, Barbara Roduner, Bill Colson, Robert Crippen, Bonnie Dunbar, Carolyn Huntoon, Charlie Walker, Dan Brandenstein, David Leestma, Duane Ross, Eileen Collins, Fani Brown Brandenburg, Frank Hughes, George Abbey, Gerald Griffin, Hoot Gibson, Jay Honeycutt, Jeff Hoffman, John Creighton, John Fabian, June Scobee Rodgers, Lynn Sherr, Margaret Weitekamp, Marianne Dyson, Michael Cassutt, Michael Oldak, Mike Mullane, Rhea Seddon, Rick Hauck, Shannon Lucid, Steve Hawley, Sue Okie, Taibi Kahler, Tam O'Shaughnessy, and Wayne Hale.

While I was not able to speak with Anna Fisher directly due to her contractual obligations, I *was* able to attend her various public presentations and Q&A events, which enabled me to ask detailed questions in public forums. In particular, at Kennedy Space Center's Chat with an Astronaut event, over the course of multiple presentations, I was able to obtain Anna's answers to many of my more pressing questions. I wasn't able to speak with Kathy

Sullivan because she, too, was bound by precluding contractual obligations.

A pillar of my research was Johnson Space Center's Oral History Project, which inspired me to write this book in the first place. Owing to the tireless efforts of JSC's historians, the project now contains interviews with roughly one thousand participants, including hours of interviews with former astronauts and major NASA personnel. All of the Six, except for Judy Resnik, gave interviews for the project, and they were unreserved in their answers and tales. For that I'm grateful.

Through the Freedom of Information Act, I received from NASA audio and video footage of old interviews with the Six, as well as video of the preflight press conferences for Sally's and Judy's first missions. I also obtained a video of an in-flight press conference that Anna Fisher's crew had conducted.

I obtained transcripts and NASA documents from the University of Houston–Clear Lake's JSC archive, which holds old biographies, documents related to astronaut selection, and materials corresponding to each of the Six's inaugural flights. I also transcribed hours of space-to-ground audio from the Six's Shuttle flights, which are available on Archive.org. I pulled details from years of chronologies and press releases out of Johnson Space Center and Kennedy Space Center, including the JSC Roundups—monthly internal memos sent to NASA personnel. NASA also holds extensive press kits of the original Shuttle missions.

Tam O'Shaughnessy graciously gave me invaluable audio journals recorded by Sally Ride herself, in which she detailed her time on the Space Shuttle and her trips to Europe following her history-making mission. The National Air and Space Museum also holds many of Sally's old notes and notebooks, which proved immensely useful.

I scoured the New York Public Library's archives and obtained hundreds of newspaper articles and magazine stories about the Six, many of which included interviews with the women when they were first selected. I have more than a hundred archived TV news stories from ABC, NBC, and CBS that report on the Six's flights and accomplishments, sourced from Vanderbilt University's Television News Archive. I also pulled from the Johnny Carson archive and the *Dick Cavett Show*.

The Six is merely a snapshot of the lives of these six incredible women. To truly do justice to their stories, six separate books would be required, but I was only contracted for one. Fortunately, many rich texts appeared before mine which offer wonderful insight into these women's lives, many of which I greatly relied on for context and details.

A few of the Six themselves have taken to writing their stories. Rhea Seddon's *Go for Orbit* is a fantastic, unflinching account of her time at NASA, and I found myself thumbing through my signed copy multiple times throughout the writing process to better understand her emotions at critical points. Kathy Sullivan's *Handprints on Hubble* was also invaluable reading, since it encompasses her early life, her history-making spacewalk, and, of course, her work developing and launching the groundbreaking Hubble Space Telescope. Shannon Lucid wrote a book of her own, called *Tumbleweed*, which chronicles her training and experience living aboard the Mir space station for six months—a highlight of her career that, unfortunately, didn't fall within the arc of my book. I highly recommend these books to those who want to go deeper into the struggles—and triumphs—of the Six.

Those looking for a detailed portrait of the life of Sally Ride won't find a better one than Lynn Sherr's tremendous *Sally Ride: America's First Woman in Space*. Lynn provides a vibrant history of a woman who was notoriously private during her time on Earth, and Lynn's intrepid reporting served as an essential launching point for my book. *Sally Ride* is especially informative about Sally's storied career after leaving NASA. As Lynn and Tam point out, Sally's work as an astronaut only encompassed nine years of her life; there's much more to Sally's tale than flying to space.

Since Judy Resnik was such a private person, detailed accounts of her life are relatively scarce. However, there are a few remarkable texts I relied on to get insight into her personality and career. *Esquire* published a feature on Judy with input from her father and mother, and one of the authors of that article, Christine Spolar, wrote a chapter on Judy for the *Washington Post's* book on the *Challenger* tragedy, which honored the lives of the astronauts. Additionally, astronaut Mike Mullane, who flew with Judy on STS-41-D, wrote a hilarious and engaging memoir chronicling his time at NASA called *Riding Rockets*, in which he recounts his relationship with Judy. I turned to that book for reference several times in writing my own.

Since my book was focused on the Six, I didn't go into as much detail as I could have on the thirteen women who passed Randy Lovelace's astronaut tests. For that history, I turned to Margaret Weitekamp and her extraordinary book, *Right Stuff, Wrong Sex*, an extremely well-researched text on the history of these women and their fight to fly to space. That book goes into even greater depth on

Jackie Cochran and her involvement in both starting the program and contributing to its demise. I very much recommend it if you're looking to learn about this era.

Other texts and content I greatly relied on include Michael Cassutt's *The Astronaut Maker*, a riveting in-depth portrait of George Abbey, and David Shayler and Colin Burgess's *NASA's First Space Shuttle Astronaut Selection*, which provides a vivid account of the entire TFNG class and their integration into NASA. In writing this book I was also fortunate to be able to consult *The Real Stuff*, *Wings in Orbit*, *The Challenger Launch Decision*, *The Burning Blue*, and Netflix's *Challenger: The Final Flight* docuseries. A full list of my sourcing for *The Six* can be found on my website lorengrush.com.

THIS BOOK WOULDN'T have been possible without the open and thoughtful help of so many people. Above all, I want to thank the still-living members of the Six who participated in this project when they were able and who directed me to various sources to help tell their stories. They graciously endured my countless emails and phone calls, and the stories they shared with me and others over the years made this a dream project to work on.

I also want to thank the Six's former colleagues, notably their fellow TFNG classmates, many of whom sat with me for hours, divulging their life stories in great detail. For a few brief moments in interviews, I felt as if I was a fly on the wall during TFNG training. I want to especially thank Hoot Gibson, Steve Hawley, John Fabian, and Rick Hauck, who spoke with me for far too many hours and provided vivid and absorbing details. But I'm immensely thankful for every person who agreed to speak with me for this book.

I'm hugely grateful to Tam O'Shaughnessy for sharing with me her memories of Sally's life and providing key documents and recordings from Sally's time at NASA.

Thanks, also, to those at NASA who guided me to the right people and sourcing I needed, especially Jennifer Ross-Nazzal, Brandi Dean, Brian Odom, Robert Young, and Holly McIntyre.

A big shout-out to Paul—whose last name I do not know— at the New York Public Library. When I set out to write a book during the height of COVID-19, I naively thought it would be easy since I'd have all the time at home to work on it. But I soon learned the difficulties of trying to conduct research when every major archive in the United States was closed. Over Zoom, Paul, a researcher at the library, patiently spent hours walking me through the labyrinthian online archival system, and, thanks to his help, I was able to pull a significant amount of information without leaving my home.

I'd be nothing without the NASA chroniclers who came before me and who shared with me their insight, including Lynn Sherr, Michael Cassutt, Dave Shayler, and Stephen Slater. I only hope that my work is half as good as the stories they've told and the work they've collected over the years.

Special thanks are reserved for Robert Pearlman, who served as my space historian guide throughout this process and provided valuable advice, sourcing, and fact-checking. I'm forever in your debt and unendingly appreciative that you took all my calls.

I'm also grateful to Christine Spolar and Scott Spencer, who happily delved back into their reporting and memories from nearly thirty years ago. Thanks as well to Ryan Millager for helping to open up the past for me.

To have the time to write and focus on this book, my two successive employers—*The Verge* and *Bloomberg*—graciously gave me the time I needed to step away from full-time reporting. Not everyone has that luxury, and I consider myself very fortunate to have had that time. I also want to highlight my time at *The Verge*, a place that made me the reporter I am today and gave me the skill set I needed to write this book. My seven years there were life-changing.

To Rick Horgan, my editor, thank you for understanding the story I wanted to tell with this book and for providing the platform to do so. Together, I truly believe we've crafted something special. Susan Canavan, thank you for plucking me out of obscurity back in 2020 and taking a chance on me. You really are my fairy godmother, and you have helped change my life.

To Stephan, my mentor, there are simply not enough thank-yous to cover the amount of guidance and advice you have given me. At first, writing this book felt like wandering down a pitch-black hallway, but with your help, a light finally turned on.

Before I delved into space history, I was first and foremost a space reporter, and I've had the privilege of working alongside a truly outstanding peer group of space journalists who are the most supportive and gifted people I know. Only a tiny fraction of the population knows what it's like to do what we do, and I feel so honored to be part of this special group.

To my friends, thank you for your unwavering support and putting up with my prolonged absences on nights and weekends. Lea, Dan, and Hayley, thank you for helping me navigate the moving-picture world. And Christina, above all, I love you, and I'm so happy to have your shoulder to lean on.

Space is in my blood thanks to my parents, Gene and Joyce

Grush, who both worked for decades on the Space Shuttle program at Johnson Space Center. I want to thank them for bringing me into this world—both the Earth world and the space world—and for supporting me in all my absurd endeavors. I truly had a fantastic launchpad and I love you both so much.

And of course, I want to thank my crewmate, Chris. I could write *another* book filled with all the support and guidance you've given me during our years together. There's no one I'd rather explore this universe with.

INDEX